# Property Tables Booklet

*to accompany*

# *Thermodynamics:*
# *An Engineering Approach*

## Ninth Edition

**Yunus A. Çengel**
*University of Nevada, Reno*

**Michael A. Boles**
*North Carolina State University*

**Mehmet Kanoğlu**
*University of Gaziantep*

PROPERTY TABLES BOOKLET TO ACCOMPANY

THERMODYNAMICS: AN ENGINEERING APPROACH, NINTH EDITION

Published by McGraw-Hill Education, 2 Penn Plaza, New York, NY 10121. Copyright © 2019 by McGraw-Hill Education. All rights reserved. Printed in the United States of America. Previous edition © 2015. No part of this publication may be reproduced or distributed in any form or by any means, or stored in a database or retrieval system, without the prior written consent of McGraw-Hill Education, including, but not limited to, in any network or other electronic storage or transmission, or broadcast for distance learning.

Some ancillaries, including electronic and print components, may not be available to customers outside the United States.

This book is printed on acid-free paper.

4 5 6 7 8 LKV 23 22

ISBN 978-1-260-04899-5
MHID 1-260-04899-3

Cover image: NASA/Bill Ingalls.

mheducation.com/highered

# APPENDIX

# 1

# PROPERTY TABLES AND CHARTS
# (SI UNITS)

1

## TABLE A–1

Molar mass, gas constant, and critical-point properties

| Substance | Formula | Molar mass, $M$ kg/kmol | Gas constant, $R$ kJ/kg·K* | Critical-point properties | | |
|---|---|---|---|---|---|---|
| | | | | Temperature, K | Pressure, MPa | Volume, $m^3$/kmol |
| Air | — | 28.97 | 0.2870 | 132.5 | 3.77 | 0.0883 |
| Ammonia | $NH_3$ | 17.03 | 0.4882 | 405.5 | 11.28 | 0.0724 |
| Argon | Ar | 39.948 | 0.2081 | 151 | 4.86 | 0.0749 |
| Benzene | $C_6H_6$ | 78.115 | 0.1064 | 562 | 4.92 | 0.2603 |
| Bromine | $Br_2$ | 159.808 | 0.0520 | 584 | 10.34 | 0.1355 |
| n-Butane | $C_4H_{10}$ | 58.124 | 0.1430 | 425.2 | 3.80 | 0.2547 |
| Carbon dioxide | $CO_2$ | 44.01 | 0.1889 | 304.2 | 7.39 | 0.0943 |
| Carbon monoxide | CO | 28.011 | 0.2968 | 133 | 3.50 | 0.0930 |
| Carbon tetrachloride | $CCl_4$ | 153.82 | 0.05405 | 556.4 | 4.56 | 0.2759 |
| Chlorine | $Cl_2$ | 70.906 | 0.1173 | 417 | 7.71 | 0.1242 |
| Chloroform | $CHCl_3$ | 119.38 | 0.06964 | 536.6 | 5.47 | 0.2403 |
| Dichlorodifluoromethane (R–12) | $CCl_2F_2$ | 120.91 | 0.06876 | 384.7 | 4.01 | 0.2179 |
| Dichlorofluoromethane (R–21) | $CHCl_2F$ | 102.92 | 0.08078 | 451.7 | 5.17 | 0.1973 |
| Ethane | $C_2H_6$ | 30.070 | 0.2765 | 305.5 | 4.48 | 0.1480 |
| Ethyl alcohol | $C_2H_5OH$ | 46.07 | 0.1805 | 516 | 6.38 | 0.1673 |
| Ethylene | $C_2H_4$ | 28.054 | 0.2964 | 282.4 | 5.12 | 0.1242 |
| Helium | He | 4.003 | 2.0769 | 5.3 | 0.23 | 0.0578 |
| n-Hexane | $C_6H_{14}$ | 86.179 | 0.09647 | 507.9 | 3.03 | 0.3677 |
| Hydrogen (normal) | $H_2$ | 2.016 | 4.1240 | 33.3 | 1.30 | 0.0649 |
| Krypton | Kr | 83.80 | 0.09921 | 209.4 | 5.50 | 0.0924 |
| Methane | $CH_4$ | 16.043 | 0.5182 | 191.1 | 4.64 | 0.0993 |
| Methyl alcohol | $CH_3OH$ | 32.042 | 0.2595 | 513.2 | 7.95 | 0.1180 |
| Methyl chloride | $CH_3Cl$ | 50.488 | 0.1647 | 416.3 | 6.68 | 0.1430 |
| Neon | Ne | 20.183 | 0.4119 | 44.5 | 2.73 | 0.0417 |
| Nitrogen | $N_2$ | 28.013 | 0.2968 | 126.2 | 3.39 | 0.0899 |
| Nitrous oxide | $N_2O$ | 44.013 | 0.1889 | 309.7 | 7.27 | 0.0961 |
| Oxygen | $O_2$ | 31.999 | 0.2598 | 154.8 | 5.08 | 0.0780 |
| Propane | $C_3H_8$ | 44.097 | 0.1885 | 370 | 4.26 | 0.1998 |
| Propylene | $C_3H_6$ | 42.081 | 0.1976 | 365 | 4.62 | 0.1810 |
| Sulfur dioxide | $SO_2$ | 64.063 | 0.1298 | 430.7 | 7.88 | 0.1217 |
| Tetrafluoroethane (R–134a) | $CF_3CH_2F$ | 102.03 | 0.08149 | 374.2 | 4.059 | 0.1993 |
| Trichlorofluoromethane (R–11) | $CCl_3F$ | 137.37 | 0.06052 | 471.2 | 4.38 | 0.2478 |
| Water | $H_2O$ | 18.015 | 0.4615 | 647.1 | 22.06 | 0.0560 |
| Xenon | Xe | 131.30 | 0.06332 | 289.8 | 5.88 | 0.1186 |

*The unit kJ/kg·K is equivalent to kPa·$m^3$/kg·K. The gas constant is calculated from $R = R_u/M$, where $R_u = 8.31447$ kJ/kmol·K and $M$ is the molar mass.
*Source of Data:* K. A. Kobe and R. E. Lynn, Jr., *Chemical Review* 52 (1953), pp. 117–236; and ASHRAE, *Handbook of Fundamentals* (Atlanta, GA: American Society of Heating, Refrigerating and Air-Conditioning Engineers, Inc., 1993), pp. 16.4 and 36.1.

## TABLE A–2

Ideal-gas specific heats of various common gases

*(a)* At 300 K

| Gas | Formula | Gas constant, $R$ kJ/kg·K | $c_p$ kJ/kg·K | $c_v$ kJ/kg·K | $k$ |
|---|---|---|---|---|---|
| Air | — | 0.2870 | 1.005 | 0.718 | 1.400 |
| Argon | Ar | 0.2081 | 0.5203 | 0.3122 | 1.667 |
| Butane | $C_4H_{10}$ | 0.1433 | 1.7164 | 1.5734 | 1.091 |
| Carbon dioxide | $CO_2$ | 0.1889 | 0.846 | 0.657 | 1.289 |
| Carbon monoxide | CO | 0.2968 | 1.040 | 0.744 | 1.400 |
| Ethane | $C_2H_6$ | 0.2765 | 1.7662 | 1.4897 | 1.186 |
| Ethylene | $C_2H_4$ | 0.2964 | 1.5482 | 1.2518 | 1.237 |
| Helium | He | 2.0769 | 5.1926 | 3.1156 | 1.667 |
| Hydrogen | $H_2$ | 4.1240 | 14.307 | 10.183 | 1.405 |
| Methane | $CH_4$ | 0.5182 | 2.2537 | 1.7354 | 1.299 |
| Neon | Ne | 0.4119 | 1.0299 | 0.6179 | 1.667 |
| Nitrogen | $N_2$ | 0.2968 | 1.039 | 0.743 | 1.400 |
| Octane | $C_8H_{18}$ | 0.0729 | 1.7113 | 1.6385 | 1.044 |
| Oxygen | $O_2$ | 0.2598 | 0.918 | 0.658 | 1.395 |
| Propane | $C_3H_8$ | 0.1885 | 1.6794 | 1.4909 | 1.126 |
| Steam | $H_2O$ | 0.4615 | 1.8723 | 1.4108 | 1.327 |

*Note:* The unit kJ/kg·K is equivalent to kJ/kg·°C.

*Source of Data:* B. G. Kyle, *Chemical and Process Thermodynamics, 3rd ed.* (Upper Saddle River, NJ: Prentice Hall, 2000).

## TABLE A–2

Ideal-gas specific heats of various common gases (Continued)

(b) At various temperatures

| Temperature, K | $c_p$ kJ/kg·K | $c_v$ kJ/kg·K | $k$ | $c_p$ kJ/kg·K | $c_v$ kJ/kg·K | $k$ | $c_p$ kJ/kg·K | $c_v$ kJ/kg·K | $k$ |
|---|---|---|---|---|---|---|---|---|---|
| | Air | | | Carbon dioxide, $CO_2$ | | | Carbon monoxide, $CO$ | | |
| 250 | 1.003 | 0.716 | 1.401 | 0.791 | 0.602 | 1.314 | 1.039 | 0.743 | 1.400 |
| 300 | 1.005 | 0.718 | 1.400 | 0.846 | 0.657 | 1.288 | 1.040 | 0.744 | 1.399 |
| 350 | 1.008 | 0.721 | 1.398 | 0.895 | 0.706 | 1.268 | 1.043 | 0.746 | 1.398 |
| 400 | 1.013 | 0.726 | 1.395 | 0.939 | 0.750 | 1.252 | 1.047 | 0.751 | 1.395 |
| 450 | 1.020 | 0.733 | 1.391 | 0.978 | 0.790 | 1.239 | 1.054 | 0.757 | 1.392 |
| 500 | 1.029 | 0.742 | 1.387 | 1.014 | 0.825 | 1.229 | 1.063 | 0.767 | 1.387 |
| 550 | 1.040 | 0.753 | 1.381 | 1.046 | 0.857 | 1.220 | 1.075 | 0.778 | 1.382 |
| 600 | 1.051 | 0.764 | 1.376 | 1.075 | 0.886 | 1.213 | 1.087 | 0.790 | 1.376 |
| 650 | 1.063 | 0.776 | 1.370 | 1.102 | 0.913 | 1.207 | 1.100 | 0.803 | 1.370 |
| 700 | 1.075 | 0.788 | 1.364 | 1.126 | 0.937 | 1.202 | 1.113 | 0.816 | 1.364 |
| 750 | 1.087 | 0.800 | 1.359 | 1.148 | 0.959 | 1.197 | 1.126 | 0.829 | 1.358 |
| 800 | 1.099 | 0.812 | 1.354 | 1.169 | 0.980 | 1.193 | 1.139 | 0.842 | 1.353 |
| 900 | 1.121 | 0.834 | 1.344 | 1.204 | 1.015 | 1.186 | 1.163 | 0.866 | 1.343 |
| 1000 | 1.142 | 0.855 | 1.336 | 1.234 | 1.045 | 1.181 | 1.185 | 0.888 | 1.335 |
| | Hydrogen, $H_2$ | | | Nitrogen, $N_2$ | | | Oxygen, $O_2$ | | |
| 250 | 14.051 | 9.927 | 1.416 | 1.039 | 0.742 | 1.400 | 0.913 | 0.653 | 1.398 |
| 300 | 14.307 | 10.183 | 1.405 | 1.039 | 0.743 | 1.400 | 0.918 | 0.658 | 1.395 |
| 350 | 14.427 | 10.302 | 1.400 | 1.041 | 0.744 | 1.399 | 0.928 | 0.668 | 1.389 |
| 400 | 14.476 | 10.352 | 1.398 | 1.044 | 0.747 | 1.397 | 0.941 | 0.681 | 1.382 |
| 450 | 14.501 | 10.377 | 1.398 | 1.049 | 0.752 | 1.395 | 0.956 | 0.696 | 1.373 |
| 500 | 14.513 | 10.389 | 1.397 | 1.056 | 0.759 | 1.391 | 0.972 | 0.712 | 1.365 |
| 550 | 14.530 | 10.405 | 1.396 | 1.065 | 0.768 | 1.387 | 0.988 | 0.728 | 1.358 |
| 600 | 14.546 | 10.422 | 1.396 | 1.075 | 0.778 | 1.382 | 1.003 | 0.743 | 1.350 |
| 650 | 14.571 | 10.447 | 1.395 | 1.086 | 0.789 | 1.376 | 1.017 | 0.758 | 1.343 |
| 700 | 14.604 | 10.480 | 1.394 | 1.098 | 0.801 | 1.371 | 1.031 | 0.771 | 1.337 |
| 750 | 14.645 | 10.521 | 1.392 | 1.110 | 0.813 | 1.365 | 1.043 | 0.783 | 1.332 |
| 800 | 14.695 | 10.570 | 1.390 | 1.121 | 0.825 | 1.360 | 1.054 | 0.794 | 1.327 |
| 900 | 14.822 | 10.698 | 1.385 | 1.145 | 0.849 | 1.349 | 1.074 | 0.814 | 1.319 |
| 1000 | 14.983 | 10.859 | 1.380 | 1.167 | 0.870 | 1.341 | 1.090 | 0.830 | 1.313 |

Source of Data: Kenneth Wark, Thermodynamics, 4th ed. (New York: McGraw–Hill, 1983), p. 783, Table A–4M. Originally published in Tables of Thermal Properties of Gases, NBS Circular 564, 1955.

Ideal-gas specific heats of various common gases (*Concluded*)

(c) As a function of temperature

$$\overline{c}_p = a + bT + cT^2 + dT^3$$

$$(T \text{ in K}, c_p \text{ in kJ/kmol·K})$$

| Substance | Formula | $a$ | $b$ | $c$ | $d$ | Temperature range, K | % error Max. | % error Avg. |
|---|---|---|---|---|---|---|---|---|
| Nitrogen | $N_2$ | 28.90 | $-0.1571 \times 10^{-2}$ | $0.8081 \times 10^{-5}$ | $-2.873 \times 10^{-9}$ | 273–1800 | 0.59 | 0.34 |
| Oxygen | $O_2$ | 25.48 | $1.520 \times 10^{-2}$ | $-0.7155 \times 10^{-5}$ | $1.312 \times 10^{-9}$ | 273–1800 | 1.19 | 0.28 |
| Air | — | 28.11 | $0.1967 \times 10^{-2}$ | $0.4802 \times 10^{-5}$ | $-1.966 \times 10^{-9}$ | 273–1800 | 0.72 | 0.33 |
| Hydrogen | $H_2$ | 29.11 | $-0.1916 \times 10^{-2}$ | $0.4003 \times 10^{-5}$ | $-0.8704 \times 10^{-9}$ | 273–1800 | 1.01 | 0.26 |
| Carbon monoxide | CO | 28.16 | $0.1675 \times 10^{-2}$ | $0.5372 \times 10^{-5}$ | $-2.222 \times 10^{-9}$ | 273–1800 | 0.89 | 0.37 |
| Carbon dioxide | $CO_2$ | 22.26 | $5.981 \times 10^{-2}$ | $-3.501 \times 10^{-5}$ | $7.469 \times 10^{-9}$ | 273–1800 | 0.67 | 0.22 |
| Water vapor | $H_2O$ | 32.24 | $0.1923 \times 10^{-2}$ | $1.055 \times 10^{-5}$ | $-3.595 \times 10^{-9}$ | 273–1800 | 0.53 | 0.24 |
| Nitric oxide | NO | 29.34 | $-0.09395 \times 10^{-2}$ | $0.9747 \times 10^{-5}$ | $-4.187 \times 10^{-9}$ | 273–1500 | 0.97 | 0.36 |
| Nitrous oxide | $N_2O$ | 24.11 | $5.8632 \times 10^{-2}$ | $-3.562 \times 10^{-5}$ | $10.58 \times 10^{-9}$ | 273–1500 | 0.59 | 0.26 |
| Nitrogen dioxide | $NO_2$ | 22.9 | $5.715 \times 10^{-2}$ | $-3.52 \times 10^{-5}$ | $7.87 \times 10^{-9}$ | 273–1500 | 0.46 | 0.18 |
| Ammonia | $NH_3$ | 27.568 | $2.5630 \times 10^{-2}$ | $0.99072 \times 10^{-5}$ | $-6.6909 \times 10^{-9}$ | 273–1500 | 0.91 | 0.36 |
| Sulfur | S | 27.21 | $2.218 \times 10^{-2}$ | $-1.628 \times 10^{-5}$ | $3.986 \times 10^{-9}$ | 273–1800 | 0.99 | 0.38 |
| Sulfur dioxide | $SO_2$ | 25.78 | $5.795 \times 10^{-2}$ | $-3.812 \times 10^{-5}$ | $8.612 \times 10^{-9}$ | 273–1800 | 0.45 | 0.24 |
| Sulfur trioxide | $SO_3$ | 16.40 | $14.58 \times 10^{-2}$ | $-11.20 \times 10^{-5}$ | $32.42 \times 10^{-9}$ | 273–1300 | 0.29 | 0.13 |
| Acetylene | $C_2H_2$ | 21.8 | $9.2143 \times 10^{-2}$ | $-6.527 \times 10^{-5}$ | $18.21 \times 10^{-9}$ | 273–1500 | 1.46 | 0.59 |
| Benzene | $C_6H_6$ | −36.22 | $48.475 \times 10^{-2}$ | $-31.57 \times 10^{-5}$ | $77.62 \times 10^{-9}$ | 273–1500 | 0.34 | 0.20 |
| Methanol | $CH_4O$ | 19.0 | $9.152 \times 10^{-2}$ | $-1.22 \times 10^{-5}$ | $-8.039 \times 10^{-9}$ | 273–1000 | 0.18 | 0.08 |
| Ethanol | $C_2H_6O$ | 19.9 | $20.96 \times 10^{-2}$ | $-10.38 \times 10^{-5}$ | $20.05 \times 10^{-9}$ | 273–1500 | 0.40 | 0.22 |
| Hydrogen chloride | HCl | 30.33 | $-0.7620 \times 10^{-2}$ | $1.327 \times 10^{-5}$ | $-4.338 \times 10^{-9}$ | 273–1500 | 0.22 | 0.08 |
| Methane | $CH_4$ | 19.89 | $5.024 \times 10^{-2}$ | $1.269 \times 10^{-5}$ | $-11.01 \times 10^{-9}$ | 273–1500 | 1.33 | 0.57 |
| Ethane | $C_2H_6$ | 6.900 | $17.27 \times 10^{-2}$ | $-6.406 \times 10^{-5}$ | $7.285 \times 10^{-9}$ | 273–1500 | 0.83 | 0.28 |
| Propane | $C_3H_8$ | −4.04 | $30.48 \times 10^{-2}$ | $-15.72 \times 10^{-5}$ | $31.74 \times 10^{-9}$ | 273–1500 | 0.40 | 0.12 |
| *n*-Butane | $C_4H_{10}$ | 3.96 | $37.15 \times 10^{-2}$ | $-18.34 \times 10^{-5}$ | $35.00 \times 10^{-9}$ | 273–1500 | 0.54 | 0.24 |
| *i*-Butane | $C_4H_{10}$ | −7.913 | $41.60 \times 10^{-2}$ | $-23.01 \times 10^{-5}$ | $49.91 \times 10^{-9}$ | 273–1500 | 0.25 | 0.13 |
| *n*-Pentane | $C_5H_{12}$ | 6.774 | $45.43 \times 10^{-2}$ | $-22.46 \times 10^{-5}$ | $42.29 \times 10^{-9}$ | 273–1500 | 0.56 | 0.21 |
| *n*-Hexane | $C_6H_{14}$ | 6.938 | $55.22 \times 10^{-2}$ | $-28.65 \times 10^{-5}$ | $57.69 \times 10^{-9}$ | 273–1500 | 0.72 | 0.20 |
| Ethylene | $C_2H_4$ | 3.95 | $15.64 \times 10^{-2}$ | $-8.344 \times 10^{-5}$ | $17.67 \times 10^{-9}$ | 273–1500 | 0.54 | 0.13 |
| Propylene | $C_3H_6$ | 3.15 | $23.83 \times 10^{-2}$ | $-12.18 \times 10^{-5}$ | $24.62 \times 10^{-9}$ | 273–1500 | 0.73 | 0.17 |

*Source of Data:* B. G. Kyle, *Chemical and Process Thermodynamics* (Englewood Cliffs, NJ: Prentice–Hall, 1984).

## TABLE A–3

Properties of common liquids, solids, and foods

### (a) Liquids

| Substance | Boiling data at 1 atm | | Freezing data | | Liquid properties | | |
|---|---|---|---|---|---|---|---|
| | Normal boiling point, °C | Latent heat of vaporization $h_{fg}$, kJ/kg | Freezing point, °C | Latent heat of fusion $h_{if}$, kJ/kg | Temperature, °C | Density $\rho$, kg/m$^3$ | Specific heat $c_p$, kJ/kg·K |
| Ammonia | −33.3 | 1357 | −77.7 | 322.4 | −33.3 | 682 | 4.43 |
| | | | | | −20 | 665 | 4.52 |
| | | | | | 0 | 639 | 4.60 |
| | | | | | 25 | 602 | 4.80 |
| Argon | −185.9 | 161.6 | −189.3 | 28 | −185.6 | 1394 | 1.14 |
| Benzene | 80.2 | 394 | 5.5 | 126 | 20 | 879 | 1.72 |
| Brine (20% sodium chloride by mass) | 103.9 | — | −17.4 | — | 20 | 1150 | 3.11 |
| n-Butane | −0.5 | 385.2 | −138.5 | 80.3 | −0.5 | 601 | 2.31 |
| Carbon dioxide | −78.4* | 230.5 (at 0°C) | −56.6 | | 0 | 298 | 0.59 |
| Ethanol | 78.2 | 838.3 | −114.2 | 109 | 25 | 783 | 2.46 |
| Ethyl alcohol | 78.6 | 855 | −156 | 108 | 20 | 789 | 2.84 |
| Ethylene glycol | 198.1 | 800.1 | −10.8 | 181.1 | 20 | 1109 | 2.84 |
| Glycerine | 179.9 | 974 | 18.9 | 200.6 | 20 | 1261 | 2.32 |
| Helium | −268.9 | 22.8 | — | — | −268.9 | 146.2 | 22.8 |
| Hydrogen | −252.8 | 445.7 | −259.2 | 59.5 | −252.8 | 70.7 | 10.0 |
| Isobutane | −11.7 | 367.1 | −160 | 105.7 | −11.7 | 593.8 | 2.28 |
| Kerosene | 204–293 | 251 | −24.9 | — | 20 | 820 | 2.00 |
| Mercury | 356.7 | 294.7 | −38.9 | 11.4 | 25 | 13,560 | 0.139 |
| Methane | −161.5 | 510.4 | −182.2 | 58.4 | −161.5 | 423 | 3.49 |
| | | | | | −100 | 301 | 5.79 |
| Methanol | 64.5 | 1100 | −97.7 | 99.2 | 25 | 787 | 2.55 |
| Nitrogen | −195.8 | 198.6 | −210 | 25.3 | −195.8 | 809 | 2.06 |
| | | | | | −160 | 596 | 2.97 |
| Octane | 124.8 | 306.3 | −57.5 | 180.7 | 20 | 703 | 2.10 |
| Oil (light) | | | | | 25 | 910 | 1.80 |
| Oxygen | −183 | 212.7 | −218.8 | 13.7 | −183 | 1141 | 1.71 |
| Petroleum | — | 230–384 | | | 20 | 640 | 2.0 |
| Propane | −42.1 | 427.8 | −187.7 | 80.0 | −42.1 | 581 | 2.25 |
| | | | | | 0 | 529 | 2.53 |
| | | | | | 50 | 449 | 3.13 |
| Refrigerant-134a | −26.1 | 217.0 | −96.6 | — | −50 | 1443 | 1.23 |
| | | | | | −26.1 | 1374 | 1.27 |
| | | | | | 0 | 1295 | 1.34 |
| | | | | | 25 | 1207 | 1.43 |
| Water | 100 | 2257 | 0.0 | 333.7 | 0 | 1000 | 4.22 |
| | | | | | 25 | 997 | 4.18 |
| | | | | | 50 | 988 | 4.18 |
| | | | | | 75 | 975 | 4.19 |
| | | | | | 100 | 958 | 4.22 |

*Sublimation temperature. (At pressures below the triple–point pressure of 518 kPa, carbon dioxide exists as a solid or gas. Also, the freezing–point temperature of carbon dioxide is the triple–point temperature of −56.5°C.)

## TABLE A–3

Properties of common liquids, solids, and foods (*Concluded*)

### (b) Solids (values are for room temperature unless indicated otherwise)

| Substance | Density, $\rho$ kg/m$^3$ | Specific heat, $c_p$ kJ/kg·K | Substance | Density, $\rho$ kg/m$^3$ | Specific heat, $c_p$ kJ/kg·K |
|---|---|---|---|---|---|
| **Metals** | | | **Nonmetals** | | |
| Aluminum | | | Asphalt | 2110 | 0.920 |
| 200 K | | 0.797 | Brick, common | 1922 | 0.79 |
| 250 K | | 0.859 | Brick, fireclay (500°C) | 2300 | 0.960 |
| 300 K | 2,700 | 0.902 | Concrete | 2300 | 0.653 |
| 350 K | | 0.929 | Clay | 1000 | 0.920 |
| 400 K | | 0.949 | Diamond | 2420 | 0.616 |
| 450 K | | 0.973 | Glass, window | 2700 | 0.800 |
| 500 K | | 0.997 | Glass, pyrex | 2230 | 0.840 |
| Bronze (76% Cu, 2% Zn, 2% Al) | 8,280 | 0.400 | Graphite | 2500 | 0.711 |
| | | | Granite | 2700 | 1.017 |
| Brass, yellow (65% Cu, 35% Zn) | 8,310 | 0.400 | Gypsum or plaster board | 800 | 1.09 |
| | | | Ice | | |
| Copper | | | 200 K | | 1.56 |
| −173°C | | 0.254 | 220 K | | 1.71 |
| −100°C | | 0.342 | 240 K | | 1.86 |
| −50°C | | 0.367 | 260 K | | 2.01 |
| 0°C | | 0.381 | 273 K | 921 | 2.11 |
| 27°C | 8,900 | 0.386 | Limestone | 1650 | 0.909 |
| 100°C | | 0.393 | Marble | 2600 | 0.880 |
| 200°C | | 0.403 | Plywood (Douglas Fir) | 545 | 1.21 |
| Iron | 7,840 | 0.45 | Rubber (soft) | 1100 | 1.840 |
| Lead | 11,310 | 0.128 | Rubber (hard) | 1150 | 2.009 |
| Magnesium | 1,730 | 1.000 | Sand | 1520 | 0.800 |
| Nickel | 8,890 | 0.440 | Stone | 1500 | 0.800 |
| Silver | 10,470 | 0.235 | Woods, hard (maple, oak, etc.) | 721 | 1.26 |
| Steel, mild | 7,830 | 0.500 | Woods, soft (fir, pine, etc.) | 513 | 1.38 |
| Tungsten | 19,400 | 0.130 | | | |

### (c) Foods

| Food | Water content, % (mass) | Freezing point, °C | Specific heat, kJ/kg·K Above freezing | Specific heat, kJ/kg·K Below freezing | Latent heat of fusion, kJ/kg | Food | Water content, % (mass) | Freezing point, °C | Specific heat, kJ/kg·K Above freezing | Specific heat, kJ/kg·K Below freezing | Latent heat of fusion, kJ/kg |
|---|---|---|---|---|---|---|---|---|---|---|---|
| Apples | 84 | −1.1 | 3.65 | 1.90 | 281 | Lettuce | 95 | −0.2 | 4.02 | 2.04 | 317 |
| Bananas | 75 | −0.8 | 3.35 | 1.78 | 251 | Milk, whole | 88 | −0.6 | 3.79 | 1.95 | 294 |
| Beef round | 67 | — | 3.08 | 1.68 | 224 | Oranges | 87 | −0.8 | 3.75 | 1.94 | 291 |
| Broccoli | 90 | −0.6 | 3.86 | 1.97 | 301 | Potatoes | 78 | −0.6 | 3.45 | 1.82 | 261 |
| Butter | 16 | — | — | 1.04 | 53 | Salmon fish | 64 | −2.2 | 2.98 | 1.65 | 214 |
| Cheese, swiss | 39 | −10.0 | 2.15 | 1.33 | 130 | Shrimp | 83 | −2.2 | 3.62 | 1.89 | 277 |
| Cherries | 80 | −1.8 | 3.52 | 1.85 | 267 | Spinach | 93 | −0.3 | 3.96 | 2.01 | 311 |
| Chicken | 74 | −2.8 | 3.32 | 1.77 | 247 | Strawberries | 90 | −0.8 | 3.86 | 1.97 | 301 |
| Corn, sweet | 74 | −0.6 | 3.32 | 1.77 | 247 | Tomatoes, ripe | 94 | −0.5 | 3.99 | 2.02 | 314 |
| Eggs, whole | 74 | −0.6 | 3.32 | 1.77 | 247 | Turkey | 64 | — | 2.98 | 1.65 | 214 |
| Ice cream | 63 | −5.6 | 2.95 | 1.63 | 210 | Watermelon | 93 | −0.4 | 3.96 | 2.01 | 311 |

*Source of Data:* Values are obtained from various handbooks and other sources or are calculated. Water content and freezing–point data of foods are from *ASHRAE, Handbook of Fundamentals,* SI version (Atlanta, GA: American Society of Heating, Refrigerating and Air-Conditioning Engineers, Inc., 1993), Chapter 30, Table1. Freezing point is the temperature at which freezing starts for fruits and vegetables, and the average freezing temperature for other foods.

## TABLE A–4

Saturated water—Temperature table

| Temp., $T$ °C | Sat. Press., $P_{sat}$ kPa | Specific volume, m³/kg | | Internal energy, kJ/kg | | | Enthalpy, kJ/kg | | | Entropy, kJ/kg·K | | |
|---|---|---|---|---|---|---|---|---|---|---|---|---|
| | | Sat. liquid, $v_f$ | Sat. vapor, $v_g$ | Sat. liquid, $u_f$ | Evap., $u_{fg}$ | Sat. vapor, $u_g$ | Sat. liquid, $h_f$ | Evap., $h_{fg}$ | Sat. vapor, $h_g$ | Sat. liquid, $s_f$ | Evap., $s_{fg}$ | Sat. vapor, $s_g$ |
| 0.01 | 0.6117 | 0.001000 | 206.00 | 0.000 | 2374.9 | 2374.9 | 0.001 | 2500.9 | 2500.9 | 0.0000 | 9.1556 | 9.1556 |
| 5 | 0.8725 | 0.001000 | 147.03 | 21.019 | 2360.8 | 2381.8 | 21.020 | 2489.1 | 2510.1 | 0.0763 | 8.9487 | 9.0249 |
| 10 | 1.2281 | 0.001000 | 106.32 | 42.020 | 2346.6 | 2388.7 | 42.022 | 2477.2 | 2519.2 | 0.1511 | 8.7488 | 8.8999 |
| 15 | 1.7057 | 0.001001 | 77.885 | 62.980 | 2332.5 | 2395.5 | 62.982 | 2465.4 | 2528.3 | 0.2245 | 8.5559 | 8.7803 |
| 20 | 2.3392 | 0.001002 | 57.762 | 83.913 | 2318.4 | 2402.3 | 83.915 | 2453.5 | 2537.4 | 0.2965 | 8.3696 | 8.6661 |
| 25 | 3.1698 | 0.001003 | 43.340 | 104.83 | 2304.3 | 2409.1 | 104.83 | 2441.7 | 2546.5 | 0.3672 | 8.1895 | 8.5567 |
| 30 | 4.2469 | 0.001004 | 32.879 | 125.73 | 2290.2 | 2415.9 | 125.74 | 2429.8 | 2555.6 | 0.4368 | 8.0152 | 8.4520 |
| 35 | 5.6291 | 0.001006 | 25.205 | 146.63 | 2276.0 | 2422.7 | 146.64 | 2417.9 | 2564.6 | 0.5051 | 7.8466 | 8.3517 |
| 40 | 7.3851 | 0.001008 | 19.515 | 167.53 | 2261.9 | 2429.4 | 167.53 | 2406.0 | 2573.5 | 0.5724 | 7.6832 | 8.2556 |
| 45 | 9.5953 | 0.001010 | 15.251 | 188.43 | 2247.7 | 2436.1 | 188.44 | 2394.0 | 2582.4 | 0.6386 | 7.5247 | 8.1633 |
| 50 | 12.352 | 0.001012 | 12.026 | 209.33 | 2233.4 | 2442.7 | 209.34 | 2382.0 | 2591.3 | 0.7038 | 7.3710 | 8.0748 |
| 55 | 15.763 | 0.001015 | 9.5639 | 230.24 | 2219.1 | 2449.3 | 230.26 | 2369.8 | 2600.1 | 0.7680 | 7.2218 | 7.9898 |
| 60 | 19.947 | 0.001017 | 7.6670 | 251.16 | 2204.7 | 2455.9 | 251.18 | 2357.7 | 2608.8 | 0.8313 | 7.0769 | 7.9082 |
| 65 | 25.043 | 0.001020 | 6.1935 | 272.09 | 2190.3 | 2462.4 | 272.12 | 2345.4 | 2617.5 | 0.8937 | 6.9360 | 7.8296 |
| 70 | 31.202 | 0.001023 | 5.0396 | 293.04 | 2175.8 | 2468.9 | 293.07 | 2333.0 | 2626.1 | 0.9551 | 6.7989 | 7.7540 |
| 75 | 38.597 | 0.001026 | 4.1291 | 313.99 | 2161.3 | 2475.3 | 314.03 | 2320.6 | 2634.6 | 1.0158 | 6.6655 | 7.6812 |
| 80 | 47.416 | 0.001029 | 3.4053 | 334.97 | 2146.6 | 2481.6 | 335.02 | 2308.0 | 2643.0 | 1.0756 | 6.5355 | 7.6111 |
| 85 | 57.868 | 0.001032 | 2.8261 | 355.96 | 2131.9 | 2487.8 | 356.02 | 2295.3 | 2651.4 | 1.1346 | 6.4089 | 7.5435 |
| 90 | 70.183 | 0.001036 | 2.3593 | 376.97 | 2117.0 | 2494.0 | 377.04 | 2282.5 | 2659.6 | 1.1929 | 6.2853 | 7.4782 |
| 95 | 84.609 | 0.001040 | 1.9808 | 398.00 | 2102.0 | 2500.1 | 398.09 | 2269.6 | 2667.6 | 1.2504 | 6.1647 | 7.4151 |
| 100 | 101.42 | 0.001043 | 1.6720 | 419.06 | 2087.0 | 2506.0 | 419.17 | 2256.4 | 2675.6 | 1.3072 | 6.0470 | 7.3542 |
| 105 | 120.90 | 0.001047 | 1.4186 | 440.15 | 2071.8 | 2511.9 | 440.28 | 2243.1 | 2683.4 | 1.3634 | 5.9319 | 7.2952 |
| 110 | 143.38 | 0.001052 | 1.2094 | 461.27 | 2056.4 | 2517.7 | 461.42 | 2229.7 | 2691.1 | 1.4188 | 5.8193 | 7.2382 |
| 115 | 169.18 | 0.001056 | 1.0360 | 482.42 | 2040.9 | 2523.3 | 482.59 | 2216.0 | 2698.6 | 1.4737 | 5.7092 | 7.1829 |
| 120 | 198.67 | 0.001060 | 0.89133 | 503.60 | 2025.3 | 2528.9 | 503.81 | 2202.1 | 2706.0 | 1.5279 | 5.6013 | 7.1292 |
| 125 | 232.23 | 0.001065 | 0.77012 | 524.83 | 2009.5 | 2534.3 | 525.07 | 2188.1 | 2713.1 | 1.5816 | 5.4956 | 7.0771 |
| 130 | 270.28 | 0.001070 | 0.66808 | 546.10 | 1993.4 | 2539.5 | 546.38 | 2173.7 | 2720.1 | 1.6346 | 5.3919 | 7.0265 |
| 135 | 313.22 | 0.001075 | 0.58179 | 567.41 | 1977.3 | 2544.7 | 567.75 | 2159.1 | 2726.9 | 1.6872 | 5.2901 | 6.9773 |
| 140 | 361.53 | 0.001080 | 0.50850 | 588.77 | 1960.9 | 2549.6 | 589.16 | 2144.3 | 2733.5 | 1.7392 | 5.1901 | 6.9294 |
| 145 | 415.68 | 0.001085 | 0.44600 | 610.19 | 1944.2 | 2554.4 | 610.64 | 2129.2 | 2739.8 | 1.7908 | 5.0919 | 6.8827 |
| 150 | 476.16 | 0.001091 | 0.39248 | 631.66 | 1927.4 | 2559.1 | 632.18 | 2113.8 | 2745.9 | 1.8418 | 4.9953 | 6.8371 |
| 155 | 543.49 | 0.001096 | 0.34648 | 653.19 | 1910.3 | 2563.5 | 653.79 | 2098.0 | 2751.8 | 1.8924 | 4.9002 | 6.7927 |
| 160 | 618.23 | 0.001102 | 0.30680 | 674.79 | 1893.0 | 2567.8 | 675.47 | 2082.0 | 2757.5 | 1.9426 | 4.8066 | 6.7492 |
| 165 | 700.93 | 0.001108 | 0.27244 | 696.46 | 1875.4 | 2571.9 | 697.24 | 2065.6 | 2762.8 | 1.9923 | 4.7143 | 6.7067 |
| 170 | 792.18 | 0.001114 | 0.24260 | 718.20 | 1857.5 | 2575.7 | 719.08 | 2048.8 | 2767.9 | 2.0417 | 4.6233 | 6.6650 |
| 175 | 892.60 | 0.001121 | 0.21659 | 740.02 | 1839.4 | 2579.4 | 741.02 | 2031.7 | 2772.7 | 2.0906 | 4.5335 | 6.6242 |
| 180 | 1002.8 | 0.001127 | 0.19384 | 761.92 | 1820.9 | 2582.8 | 763.05 | 2014.2 | 2777.2 | 2.1392 | 4.4448 | 6.5841 |
| 185 | 1123.5 | 0.001134 | 0.17390 | 783.91 | 1802.1 | 2586.0 | 785.19 | 1996.2 | 2781.4 | 2.1875 | 4.3572 | 6.5447 |
| 190 | 1255.2 | 0.001141 | 0.15636 | 806.00 | 1783.0 | 2589.0 | 807.43 | 1977.9 | 2785.3 | 2.2355 | 4.2705 | 6.5059 |
| 195 | 1398.8 | 0.001149 | 0.14089 | 828.18 | 1763.6 | 2591.7 | 829.78 | 1959.0 | 2788.8 | 2.2831 | 4.1847 | 6.4678 |
| 200 | 1554.9 | 0.001157 | 0.12721 | 850.46 | 1743.7 | 2594.2 | 852.26 | 1939.8 | 2792.0 | 2.3305 | 4.0997 | 6.4302 |

## TABLE A–4

Saturated water—Temperature table (Concluded)

| Temp., $T\,°C$ | Sat. Press., $P_{sat}$ kPa | Specific volume, m³/kg | | Internal energy, kJ/kg | | | Enthalpy, kJ/kg | | | Entropy, kJ/kg·K | | |
|---|---|---|---|---|---|---|---|---|---|---|---|---|
| | | Sat. liquid, $v_f$ | Sat. vapor, $v_g$ | Sat. liquid, $u_f$ | Evap., $u_{fg}$ | Sat. vapor, $u_g$ | Sat. liquid, $h_f$ | Evap., $h_{fg}$ | Sat. vapor, $h_g$ | Sat. liquid, $s_f$ | Evap., $s_{fg}$ | Sat. vapor, $s_g$ |
| 205 | 1724.3 | 0.001164 | 0.11508 | 872.86 | 1723.5 | 2596.4 | 874.87 | 1920.0 | 2794.8 | 2.3776 | 4.0154 | 6.3930 |
| 210 | 1907.7 | 0.001173 | 0.10429 | 895.38 | 1702.9 | 2598.3 | 897.61 | 1899.7 | 2797.3 | 2.4245 | 3.9318 | 6.3563 |
| 215 | 2105.9 | 0.001181 | 0.094680 | 918.02 | 1681.9 | 2599.9 | 920.50 | 1878.8 | 2799.3 | 2.4712 | 3.8489 | 6.3200 |
| 220 | 2319.6 | 0.001190 | 0.086094 | 940.79 | 1660.5 | 2601.3 | 943.55 | 1857.4 | 2801.0 | 2.5176 | 3.7664 | 6.2840 |
| 225 | 2549.7 | 0.001199 | 0.078405 | 963.70 | 1638.6 | 2602.3 | 966.76 | 1835.4 | 2802.2 | 2.5639 | 3.6844 | 6.2483 |
| 230 | 2797.1 | 0.001209 | 0.071505 | 986.76 | 1616.1 | 2602.9 | 990.14 | 1812.8 | 2802.9 | 2.6100 | 3.6028 | 6.2128 |
| 235 | 3062.6 | 0.001219 | 0.065300 | 1010.0 | 1593.2 | 2603.2 | 1013.7 | 1789.5 | 2803.2 | 2.6560 | 3.5216 | 6.1775 |
| 240 | 3347.0 | 0.001229 | 0.059707 | 1033.4 | 1569.8 | 2603.1 | 1037.5 | 1765.5 | 2803.0 | 2.7018 | 3.4405 | 6.1424 |
| 245 | 3651.2 | 0.001240 | 0.054656 | 1056.9 | 1545.7 | 2602.7 | 1061.5 | 1740.8 | 2802.2 | 2.7476 | 3.3596 | 6.1072 |
| 250 | 3976.2 | 0.001252 | 0.050085 | 1080.7 | 1521.1 | 2601.8 | 1085.7 | 1715.3 | 2801.0 | 2.7933 | 3.2788 | 6.0721 |
| 255 | 4322.9 | 0.001263 | 0.045941 | 1104.7 | 1495.8 | 2600.5 | 1110.1 | 1689.0 | 2799.1 | 2.8390 | 3.1979 | 6.0369 |
| 260 | 4692.3 | 0.001276 | 0.042175 | 1128.8 | 1469.9 | 2598.7 | 1134.8 | 1661.8 | 2796.6 | 2.8847 | 3.1169 | 6.0017 |
| 265 | 5085.3 | 0.001289 | 0.038748 | 1153.3 | 1443.2 | 2596.5 | 1159.8 | 1633.7 | 2793.5 | 2.9304 | 3.0358 | 5.9662 |
| 270 | 5503.0 | 0.001303 | 0.035622 | 1177.9 | 1415.7 | 2593.7 | 1185.1 | 1604.6 | 2789.7 | 2.9762 | 2.9542 | 5.9305 |
| 275 | 5946.4 | 0.001317 | 0.032767 | 1202.9 | 1387.4 | 2590.3 | 1210.7 | 1574.5 | 2785.2 | 3.0221 | 2.8723 | 5.8944 |
| 280 | 6416.6 | 0.001333 | 0.030153 | 1228.2 | 1358.2 | 2586.4 | 1236.7 | 1543.2 | 2779.9 | 3.0681 | 2.7898 | 5.8579 |
| 285 | 6914.6 | 0.001349 | 0.027756 | 1253.7 | 1328.1 | 2581.8 | 1263.1 | 1510.7 | 2773.7 | 3.1144 | 2.7066 | 5.8210 |
| 290 | 7441.8 | 0.001366 | 0.025554 | 1279.7 | 1296.9 | 2576.5 | 1289.8 | 1476.9 | 2766.7 | 3.1608 | 2.6225 | 5.7834 |
| 295 | 7999.0 | 0.001384 | 0.023528 | 1306.0 | 1264.5 | 2570.5 | 1317.1 | 1441.6 | 2758.7 | 3.2076 | 2.5374 | 5.7450 |
| 300 | 8587.9 | 0.001404 | 0.021659 | 1332.7 | 1230.9 | 2563.6 | 1344.8 | 1404.8 | 2749.6 | 3.2548 | 2.4511 | 5.7059 |
| 305 | 9209.4 | 0.001425 | 0.019932 | 1360.0 | 1195.9 | 2555.8 | 1373.1 | 1366.3 | 2739.4 | 3.3024 | 2.3633 | 5.6657 |
| 310 | 9865.0 | 0.001447 | 0.018333 | 1387.7 | 1159.3 | 2547.1 | 1402.0 | 1325.9 | 2727.9 | 3.3506 | 2.2737 | 5.6243 |
| 315 | 10,556 | 0.001472 | 0.016849 | 1416.1 | 1121.1 | 2537.2 | 1431.6 | 1283.4 | 2715.0 | 3.3994 | 2.1821 | 5.5816 |
| 320 | 11,284 | 0.001499 | 0.015470 | 1445.1 | 1080.9 | 2526.0 | 1462.0 | 1238.5 | 2700.6 | 3.4491 | 2.0881 | 5.5372 |
| 325 | 12,051 | 0.001528 | 0.014183 | 1475.0 | 1038.5 | 2513.4 | 1493.4 | 1191.0 | 2684.3 | 3.4998 | 1.9911 | 5.4908 |
| 330 | 12,858 | 0.001560 | 0.012979 | 1505.7 | 993.5 | 2499.2 | 1525.8 | 1140.3 | 2666.0 | 3.5516 | 1.8906 | 5.4422 |
| 335 | 13,707 | 0.001597 | 0.011848 | 1537.5 | 945.5 | 2483.0 | 1559.4 | 1086.0 | 2645.4 | 3.6050 | 1.7857 | 5.3907 |
| 340 | 14,601 | 0.001638 | 0.010783 | 1570.7 | 893.8 | 2464.5 | 1594.6 | 1027.4 | 2622.0 | 3.6602 | 1.6756 | 5.3358 |
| 345 | 15,541 | 0.001685 | 0.009772 | 1605.5 | 837.7 | 2443.2 | 1631.7 | 963.4 | 2595.1 | 3.7179 | 1.5585 | 5.2765 |
| 350 | 16,529 | 0.001741 | 0.008806 | 1642.4 | 775.9 | 2418.3 | 1671.2 | 892.7 | 2563.9 | 3.7788 | 1.4326 | 5.2114 |
| 355 | 17,570 | 0.001808 | 0.007872 | 1682.2 | 706.4 | 2388.6 | 1714.0 | 812.9 | 2526.9 | 3.8442 | 1.2942 | 5.1384 |
| 360 | 18,666 | 0.001895 | 0.006950 | 1726.2 | 625.7 | 2351.9 | 1761.5 | 720.1 | 2481.6 | 3.9165 | 1.1373 | 5.0537 |
| 365 | 19,822 | 0.002015 | 0.006009 | 1777.2 | 526.4 | 2303.6 | 1817.2 | 605.5 | 2422.7 | 4.0004 | 0.9489 | 4.9493 |
| 370 | 21,044 | 0.002217 | 0.004953 | 1844.5 | 385.6 | 2230.1 | 1891.2 | 443.1 | 2334.3 | 4.1119 | 0.6890 | 4.8009 |
| 373.95 | 22,064 | 0.003106 | 0.003106 | 2015.7 | 0 | 2015.7 | 2084.3 | 0 | 2084.3 | 4.4070 | 0 | 4.4070 |

*Source of Data:* Tables A–4 through A–8 are generated using the Engineering Equation Solver (EES) software developed by S. A. Klein and F. L. Alvarado. The routine used in calculations is the highly accurate Steam_IAPWS, which incorporates the 1995 Formulation for the Thermodynamic Properties of Ordinary Water Substance for General and Scientific Use, issued by The International Association for the Properties of Water and Steam (IAPWS). This formulation replaces the 1984 formulation of Haar, Gallagher, and Kell (*NBS/NRC Steam Tables,* Hemisphere Publishing Co., 1984), which is also available in EES as the routine STEAM. The new formulation is based on the correlations of Saul and Wagner (*J. Phys. Chem. Ref. Data,* 16, 893, 1987) with modifications to adjust to the International Temperature Scale of 1990. The modifications are described by Wagner and Pruss (*J. Phys. Chem. Ref. Data,* 22, 783, 1993). The properties of ice are based on Hyland and Wexler, "Formulations for the Thermodynamic Properties of the Saturated Phases of H₂O from 173.15 K to 473.15 K," *ASHRAE Trans.*, Part 2A, Paper 2793, 1983.

## TABLE A–5

Saturated water—Pressure table

| Press., $P$ kPa | Sat. temp., $T_{sat}$ °C | Specific volume, m³/kg | | Internal energy, kJ/kg | | | Enthalpy, kJ/kg | | | Entropy, kJ/kg·K | | |
|---|---|---|---|---|---|---|---|---|---|---|---|---|
| | | Sat. liquid, $v_f$ | Sat. vapor, $v_g$ | Sat. liquid, $u_f$ | Evap., $u_{fg}$ | Sat. vapor, $u_g$ | Sat. liquid, $h_f$ | Evap., $h_{fg}$ | Sat. vapor, $h_g$ | Sat. liquid, $s_f$ | Evap., $s_{fg}$ | Sat. vapor, $s_g$ |
| 1.0 | 6.97 | 0.001000 | 129.19 | 29.302 | 2355.2 | 2384.5 | 29.303 | 2484.4 | 2513.7 | 0.1059 | 8.8690 | 8.9749 |
| 1.5 | 13.02 | 0.001001 | 87.964 | 54.686 | 2338.1 | 2392.8 | 54.688 | 2470.1 | 2524.7 | 0.1956 | 8.6314 | 8.8270 |
| 2.0 | 17.50 | 0.001001 | 66.990 | 73.431 | 2325.5 | 2398.9 | 73.433 | 2459.5 | 2532.9 | 0.2606 | 8.4621 | 8.7227 |
| 2.5 | 21.08 | 0.001002 | 54.242 | 88.422 | 2315.4 | 2403.8 | 88.424 | 2451.0 | 2539.4 | 0.3118 | 8.3302 | 8.6421 |
| 3.0 | 24.08 | 0.001003 | 45.654 | 100.98 | 2306.9 | 2407.9 | 100.98 | 2443.9 | 2544.8 | 0.3543 | 8.2222 | 8.5765 |
| 4.0 | 28.96 | 0.001004 | 34.791 | 121.39 | 2293.1 | 2414.5 | 121.39 | 2432.3 | 2553.7 | 0.4224 | 8.0510 | 8.4734 |
| 5.0 | 32.87 | 0.001005 | 28.185 | 137.75 | 2282.1 | 2419.8 | 137.75 | 2423.0 | 2560.7 | 0.4762 | 7.9176 | 8.3938 |
| 7.5 | 40.29 | 0.001008 | 19.233 | 168.74 | 2261.1 | 2429.8 | 168.75 | 2405.3 | 2574.0 | 0.5763 | 7.6738 | 8.2501 |
| 10 | 45.81 | 0.001010 | 14.670 | 191.79 | 2245.4 | 2437.2 | 191.81 | 2392.1 | 2583.9 | 0.6492 | 7.4996 | 8.1488 |
| 15 | 53.97 | 0.001014 | 10.020 | 225.93 | 2222.1 | 2448.0 | 225.94 | 2372.3 | 2598.3 | 0.7549 | 7.2522 | 8.0071 |
| 20 | 60.06 | 0.001017 | 7.6481 | 251.40 | 2204.6 | 2456.0 | 251.42 | 2357.5 | 2608.9 | 0.8320 | 7.0752 | 7.9073 |
| 25 | 64.96 | 0.001020 | 6.2034 | 271.93 | 2190.4 | 2462.4 | 271.96 | 2345.5 | 2617.5 | 0.8932 | 6.9370 | 7.8302 |
| 30 | 69.09 | 0.001022 | 5.2287 | 289.24 | 2178.5 | 2467.7 | 289.27 | 2335.3 | 2624.6 | 0.9441 | 6.8234 | 7.7675 |
| 40 | 75.86 | 0.001026 | 3.9933 | 317.58 | 2158.8 | 2476.3 | 317.62 | 2318.4 | 2636.1 | 1.0261 | 6.6430 | 7.6691 |
| 50 | 81.32 | 0.001030 | 3.2403 | 340.49 | 2142.7 | 2483.2 | 340.54 | 2304.7 | 2645.2 | 1.0912 | 6.5019 | 7.5931 |
| 75 | 91.76 | 0.001037 | 2.2172 | 384.36 | 2111.8 | 2496.1 | 384.44 | 2278.0 | 2662.4 | 1.2132 | 6.2426 | 7.4558 |
| 100 | 99.61 | 0.001043 | 1.6941 | 417.40 | 2088.2 | 2505.6 | 417.51 | 2257.5 | 2675.0 | 1.3028 | 6.0562 | 7.3589 |
| 101.325 | 99.97 | 0.001043 | 1.6734 | 418.95 | 2087.0 | 2506.0 | 419.06 | 2256.5 | 2675.6 | 1.3069 | 6.0476 | 7.3545 |
| 125 | 105.97 | 0.001048 | 1.3750 | 444.23 | 2068.8 | 2513.0 | 444.36 | 2240.6 | 2684.9 | 1.3741 | 5.9100 | 7.2841 |
| 150 | 111.35 | 0.001053 | 1.1594 | 466.97 | 2052.3 | 2519.2 | 467.13 | 2226.0 | 2693.1 | 1.4337 | 5.7894 | 7.2231 |
| 175 | 116.04 | 0.001057 | 1.0037 | 486.82 | 2037.7 | 2524.5 | ′ 487.01 | 2213.1 | 2700.2 | 1.4850 | 5.6865 | 7.1716 |
| 200 | 120.21 | 0.001061 | 0.88578 | 504.50 | 2024.6 | 2529.1 | 504.71 | 2201.6 | 2706.3 | 1.5302 | 5.5968 | 7.1270 |
| 225 | 123.97 | 0.001064 | 0.79329 | 520.47 | 2012.7 | 2533.2 | 520.71 | 2191.0 | 2711.7 | 1.5706 | 5.5171 | 7.0877 |
| 250 | 127.41 | 0.001067 | 0.71873 | 535.08 | 2001.8 | 2536.8 | 535.35 | 2181.2 | 2716.5 | 1.6072 | 5.4453 | 7.0525 |
| 275 | 130.58 | 0.001070 | 0.65732 | 548.57 | 1991.6 | 2540.1 | 548.86 | 2172.0 | 2720.9 | 1.6408 | 5.3800 | 7.0207 |
| 300 | 133.52 | 0.001073 | 0.60582 | 561.11 | 1982.1 | 2543.2 | 561.43 | 2163.5 | 2724.9 | 1.6717 | 5.3200 | 6.9917 |
| 325 | 136.27 | 0.001076 | 0.56199 | 572.84 | 1973.1 | 2545.9 | 573.19 | 2155.4 | 2728.6 | 1.7005 | 5.2645 | 6.9650 |
| 350 | 138.86 | 0.001079 | 0.52422 | 583.89 | 1964.6 | 2548.5 | 584.26 | 2147.7 | 2732.0 | 1.7274 | 5.2128 | 6.9402 |
| 375 | 141.30 | 0.001081 | 0.49133 | 594.32 | 1956.6 | 2550.9 | 594.73 | 2140.4 | 2735.1 | 1.7526 | 5.1645 | 6.9171 |
| 400 | 143.61 | 0.001084 | 0.46242 | 604.22 | 1948.9 | 2553.1 | 604.66 | 2133.4 | 2738.1 | 1.7765 | 5.1191 | 6.8955 |
| 450 | 147.90 | 0.001088 | 0.41392 | 622.65 | 1934.5 | 2557.1 | 623.14 | 2120.3 | 2743.4 | 1.8205 | 5.0356 | 6.8561 |
| 500 | 151.83 | 0.001093 | 0.37483 | 639.54 | 1921.2 | 2560.7 | 640.09 | 2108.0 | 2748.1 | 1.8604 | 4.9603 | 6.8207 |
| 550 | 155.46 | 0.001097 | 0.34261 | 655.16 | 1908.8 | 2563.9 | 655.77 | 2096.6 | 2752.4 | 1.8970 | 4.8916 | 6.7886 |
| 600 | 158.83 | 0.001101 | 0.31560 | 669.72 | 1897.1 | 2566.8 | 670.38 | 2085.8 | 2756.2 | 1.9308 | 4.8285 | 6.7593 |
| 650 | 161.98 | 0.001104 | 0.29260 | 683.37 | 1886.1 | 2569.4 | 684.08 | 2075.5 | 2759.6 | 1.9623 | 4.7699 | 6.7322 |
| 700 | 164.95 | 0.001108 | 0.27278 | 696.23 | 1875.6 | 2571.8 | 697.00 | 2065.8 | 2762.8 | 1.9918 | 4.7153 | 6.7071 |
| 750 | 167.75 | 0.001111 | 0.25552 | 708.40 | 1865.6 | 2574.0 | 709.24 | 2056.4 | 2765.7 | 2.0195 | 4.6642 | 6.6837 |

## TABLE A–5

Saturated water—Pressure table (*Concluded*)

| Press., $P$ kPa | Sat. temp., $T_{sat}$ °C | Specific volume, m³/kg | | Internal energy, kJ/kg | | | Enthalpy, kJ/kg | | | Entropy, kJ/kg·K | | |
|---|---|---|---|---|---|---|---|---|---|---|---|---|
| | | Sat. liquid, $v_f$ | Sat. vapor, $v_g$ | Sat. liquid, $u_f$ | Evap., $u_{fg}$ | Sat. vapor, $u_g$ | Sat. liquid, $h_f$ | Evap., $h_{fg}$ | Sat. vapor, $h_g$ | Sat. liquid, $s_f$ | Evap., $s_{fg}$ | Sat. vapor, $s_g$ |
| 800 | 170.41 | 0.001115 | 0.24035 | 719.97 | 1856.1 | 2576.0 | 720.87 | 2047.5 | 2768.3 | 2.0457 | 4.6160 | 6.6616 |
| 850 | 172.94 | 0.001118 | 0.22690 | 731.00 | 1846.9 | 2577.9 | 731.95 | 2038.8 | 2770.8 | 2.0705 | 4.5705 | 6.6409 |
| 900 | 175.35 | 0.001121 | 0.21489 | 741.55 | 1838.1 | 2579.6 | 742.56 | 2030.5 | 2773.0 | 2.0941 | 4.5273 | 6.6213 |
| 950 | 177.66 | 0.001124 | 0.20411 | 751.67 | 1829.6 | 2581.3 | 752.74 | 2022.4 | 2775.2 | 2.1166 | 4.4862 | 6.6027 |
| 1000 | 179.88 | 0.001127 | 0.19436 | 761.39 | 1821.4 | 2582.8 | 762.51 | 2014.6 | 2777.1 | 2.1381 | 4.4470 | 6.5850 |
| 1100 | 184.06 | 0.001133 | 0.17745 | 779.78 | 1805.7 | 2585.5 | 781.03 | 1999.6 | 2780.7 | 2.1785 | 4.3735 | 6.5520 |
| 1200 | 187.96 | 0.001138 | 0.16326 | 796.96 | 1790.9 | 2587.8 | 798.33 | 1985.4 | 2783.8 | 2.2159 | 4.3058 | 6.5217 |
| 1300 | 191.60 | 0.001144 | 0.15119 | 813.10 | 1776.8 | 2589.9 | 814.59 | 1971.9 | 2786.5 | 2.2508 | 4.2428 | 6.4936 |
| 1400 | 195.04 | 0.001149 | 0.14078 | 828.35 | 1763.4 | 2591.8 | 829.96 | 1958.9 | 2788.9 | 2.2835 | 4.1840 | 6.4675 |
| 1500 | 198.29 | 0.001154 | 0.13171 | 842.82 | 1750.6 | 2593.4 | 844.55 | 1946.4 | 2791.0 | 2.3143 | 4.1287 | 6.4430 |
| 1750 | 205.72 | 0.001166 | 0.11344 | 876.12 | 1720.6 | 2596.7 | 878.16 | 1917.1 | 2795.2 | 2.3844 | 4.0033 | 6.3877 |
| 2000 | 212.38 | 0.001177 | 0.099587 | 906.12 | 1693.0 | 2599.1 | 908.47 | 1889.8 | 2798.3 | 2.4467 | 3.8923 | 6.3390 |
| 2250 | 218.41 | 0.001187 | 0.088717 | 933.54 | 1667.3 | 2600.9 | 936.21 | 1864.3 | 2800.5 | 2.5029 | 3.7926 | 6.2954 |
| 2500 | 223.95 | 0.001197 | 0.079952 | 958.87 | 1643.2 | 2602.1 | 961.87 | 1840.1 | 2801.9 | 2.5542 | 3.7016 | 6.2558 |
| 3000 | 233.85 | 0.001217 | 0.066667 | 1004.6 | 1598.5 | 2603.2 | 1008.3 | 1794.9 | 2803.2 | 2.6454 | 3.5402 | 6.1856 |
| 3500 | 242.56 | 0.001235 | 0.057061 | 1045.4 | 1557.6 | 2603.0 | 1049.7 | 1753.0 | 2802.7 | 2.7253 | 3.3991 | 6.1244 |
| 4000 | 250.35 | 0.001252 | 0.049779 | 1082.4 | 1519.3 | 2601.7 | 1087.4 | 1713.5 | 2800.8 | 2.7966 | 3.2731 | 6.0696 |
| 5000 | 263.94 | 0.001286 | 0.039448 | 1148.1 | 1448.9 | 2597.0 | 1154.5 | 1639.7 | 2794.2 | 2.9207 | 3.0530 | 5.9737 |
| 6000 | 275.59 | 0.001319 | 0.032449 | 1205.8 | 1384.1 | 2589.9 | 1213.8 | 1570.9 | 2784.6 | 3.0275 | 2.8627 | 5.8902 |
| 7000 | 285.83 | 0.001352 | 0.027378 | 1258.0 | 1323.0 | 2581.0 | 1267.5 | 1505.2 | 2772.6 | 3.1220 | 2.6927 | 5.8148 |
| 8000 | 295.01 | 0.001384 | 0.023525 | 1306.0 | 1264.5 | 2570.5 | 1317.1 | 1441.6 | 2758.7 | 3.2077 | 2.5373 | 5.7450 |
| 9000 | 303.35 | 0.001418 | 0.020489 | 1350.9 | 1207.6 | 2558.5 | 1363.7 | 1379.3 | 2742.9 | 3.2866 | 2.3925 | 5.6791 |
| 10,000 | 311.00 | 0.001452 | 0.018028 | 1393.3 | 1151.8 | 2545.2 | 1407.8 | 1317.6 | 2725.5 | 3.3603 | 2.2556 | 5.6159 |
| 11,000 | 318.08 | 0.001488 | 0.015988 | 1433.9 | 1096.6 | 2530.4 | 1450.2 | 1256.1 | 2706.3 | 3.4299 | 2.1245 | 5.5544 |
| 12,000 | 324.68 | 0.001526 | 0.014264 | 1473.0 | 1041.3 | 2514.3 | 1491.3 | 1194.1 | 2685.4 | 3.4964 | 1.9975 | 5.4939 |
| 13,000 | 330.85 | 0.001566 | 0.012781 | 1511.0 | 985.5 | 2496.6 | 1531.4 | 1131.3 | 2662.7 | 3.5606 | 1.8730 | 5.4336 |
| 14,000 | 336.67 | 0.001610 | 0.011487 | 1548.4 | 928.7 | 2477.1 | 1571.0 | 1067.0 | 2637.9 | 3.6232 | 1.7497 | 5.3728 |
| 15,000 | 342.16 | 0.001657 | 0.010341 | 1585.5 | 870.3 | 2455.7 | 1610.3 | 1000.5 | 2610.8 | 3.6848 | 1.6261 | 5.3108 |
| 16,000 | 347.36 | 0.001710 | 0.009312 | 1622.6 | 809.4 | 2432.0 | 1649.9 | 931.1 | 2581.0 | 3.7461 | 1.5005 | 5.2466 |
| 17,000 | 352.29 | 0.001770 | 0.008374 | 1660.2 | 745.1 | 2405.4 | 1690.3 | 857.4 | 2547.7 | 3.8082 | 1.3709 | 5.1791 |
| 18,000 | 356.99 | 0.001840 | 0.007504 | 1699.1 | 675.9 | 2375.0 | 1732.2 | 777.8 | 2510.0 | 3.8720 | 1.2343 | 5.1064 |
| 19,000 | 361.47 | 0.001926 | 0.006677 | 1740.3 | 598.9 | 2339.2 | 1776.8 | 689.2 | 2466.0 | 3.9396 | 1.0860 | 5.0256 |
| 20,000 | 365.75 | 0.002038 | 0.005862 | 1785.8 | 509.0 | 2294.8 | 1826.6 | 585.5 | 2412.1 | 4.0146 | 0.9164 | 4.9310 |
| 21,000 | 369.83 | 0.002207 | 0.004994 | 1841.6 | 391.9 | 2233.5 | 1888.0 | 450.4 | 2338.4 | 4.1071 | 0.7005 | 4.8076 |
| 22,000 | 373.71 | 0.002703 | 0.003644 | 1951.7 | 140.8 | 2092.4 | 2011.1 | 161.5 | 2172.6 | 4.2942 | 0.2496 | 4.5439 |
| 22,064 | 373.95 | 0.003106 | 0.003106 | 2015.7 | 0 | 2015.7 | 2084.3 | 0 | 2084.3 | 4.4070 | 0 | 4.4070 |

## TABLE A–6

Superheated water

| $T$ °C | $\upsilon$ m³/kg | $u$ kJ/kg | $h$ kJ/kg | $s$ kJ/kg·K | $\upsilon$ m³/kg | $u$ kJ/kg | $h$ kJ/kg | $s$ kJ/kg·K | $\upsilon$ m³/kg | $u$ kJ/kg | $h$ kJ/kg | $s$ kJ/kg·K |
|---|---|---|---|---|---|---|---|---|---|---|---|---|
| | $P = 0.01$ MPa (45.81°C)* | | | | $P = 0.05$ MPa (81.32°C) | | | | $P = 0.10$ MPa (99.61°C) | | | |
| Sat.† | 14.670 | 2437.2 | 2583.9 | 8.1488 | 3.2403 | 2483.2 | 2645.2 | 7.5931 | 1.6941 | 2505.6 | 2675.0 | 7.3589 |
| 50 | 14.867 | 2443.3 | 2592.0 | 8.1741 | | | | | | | | |
| 100 | 17.196 | 2515.5 | 2687.5 | 8.4489 | 3.4187 | 2511.5 | 2682.4 | 7.6953 | 1.6959 | 2506.2 | 2675.8 | 7.3611 |
| 150 | 19.513 | 2587.9 | 2783.0 | 8.6893 | 3.8897 | 2585.7 | 2780.2 | 7.9413 | 1.9367 | 2582.9 | 2776.6 | 7.6148 |
| 200 | 21.826 | 2661.4 | 2879.6 | 8.9049 | 4.3562 | 2660.0 | 2877.8 | 8.1592 | 2.1724 | 2658.2 | 2875.5 | 7.8356 |
| 250 | 24.136 | 2736.1 | 2977.5 | 9.1015 | 4.8206 | 2735.1 | 2976.2 | 8.3568 | 2.4062 | 2733.9 | 2974.5 | 8.0346 |
| 300 | 26.446 | 2812.3 | 3076.7 | 9.2827 | 5.2841 | 2811.6 | 3075.8 | 8.5387 | 2.6389 | 2810.7 | 3074.5 | 8.2172 |
| 400 | 31.063 | 2969.3 | 3280.0 | 9.6094 | 6.2094 | 2968.9 | 3279.3 | 8.8659 | 3.1027 | 2968.3 | 3278.6 | 8.5452 |
| 500 | 35.680 | 3132.9 | 3489.7 | 9.8998 | 7.1338 | 3132.6 | 3489.3 | 9.1566 | 3.5655 | 3132.2 | 3488.7 | 8.8362 |
| 600 | 40.296 | 3303.3 | 3706.3 | 10.1631 | 8.0577 | 3303.1 | 3706.0 | 9.4201 | 4.0279 | 3302.8 | 3705.6 | 9.0999 |
| 700 | 44.911 | 3480.8 | 3929.9 | 10.4056 | 8.9813 | 3480.6 | 3929.7 | 9.6626 | 4.4900 | 3480.4 | 3929.4 | 9.3424 |
| 800 | 49.527 | 3665.4 | 4160.6 | 10.6312 | 9.9047 | 3665.2 | 4160.4 | 9.8883 | 4.9519 | 3665.0 | 4160.2 | 9.5682 |
| 900 | 54.143 | 3856.9 | 4398.3 | 10.8429 | 10.8280 | 3856.8 | 4398.2 | 10.1000 | 5.4137 | 3856.7 | 4398.0 | 9.7800 |
| 1000 | 58.758 | 4055.3 | 4642.8 | 11.0429 | 11.7513 | 4055.2 | 4642.7 | 10.3000 | 5.8755 | 4055.0 | 4642.6 | 9.9800 |
| 1100 | 63.373 | 4260.0 | 4893.8 | 11.2326 | 12.6745 | 4259.9 | 4893.7 | 10.4897 | 6.3372 | 4259.8 | 4893.6 | 10.1698 |
| 1200 | 67.989 | 4470.9 | 5150.8 | 11.4132 | 13.5977 | 4470.8 | 5150.7 | 10.6704 | 6.7988 | 4470.7 | 5150.6 | 10.3504 |
| 1300 | 72.604 | 4687.4 | 5413.4 | 11.5857 | 14.5209 | 4687.3 | 5413.3 | 10.8429 | 7.2605 | 4687.2 | 5413.3 | 10.5229 |
| | $P = 0.20$ MPa (120.21°C) | | | | $P = 0.30$ MPa (133.52°C) | | | | $P = 0.40$ MPa (143.61°C) | | | |
| Sat. | 0.88578 | 2529.1 | 2706.3 | 7.1270 | 0.60582 | 2543.2 | 2724.9 | 6.9917 | 0.46242 | 2553.1 | 2738.1 | 6.8955 |
| 150 | 0.95986 | 2577.1 | 2769.1 | 7.2810 | 0.63402 | 2571.0 | 2761.2 | 7.0792 | 0.47088 | 2564.4 | 2752.8 | 6.9306 |
| 200 | 1.08049 | 2654.6 | 2870.7 | 7.5081 | 0.71643 | 2651.0 | 2865.9 | 7.3132 | 0.53434 | 2647.2 | 2860.9 | 7.1723 |
| 250 | 1.19890 | 2731.4 | 2971.2 | 7.7100 | 0.79645 | 2728.9 | 2967.9 | 7.5180 | 0.59520 | 2726.4 | 2964.5 | 7.3804 |
| 300 | 1.31623 | 2808.8 | 3072.1 | 7.8941 | 0.87535 | 2807.0 | 3069.6 | 7.7037 | 0.65489 | 2805.1 | 3067.1 | 7.5677 |
| 400 | 1.54934 | 2967.2 | 3277.0 | 8.2236 | 1.03155 | 2966.0 | 3275.5 | 8.0347 | 0.77265 | 2964.9 | 3273.9 | 7.9003 |
| 500 | 1.78142 | 3131.4 | 3487.7 | 8.5153 | 1.18672 | 3130.6 | 3486.6 | 8.3271 | 0.88936 | 3129.8 | 3485.5 | 8.1933 |
| 600 | 2.01302 | 3302.2 | 3704.8 | 8.7793 | 1.34139 | 3301.6 | 3704.0 | 8.5915 | 1.00558 | 3301.0 | 3703.3 | 8.4580 |
| 700 | 2.24434 | 3479.9 | 3928.8 | 9.0221 | 1.49580 | 3479.5 | 3928.2 | 8.8345 | 1.12152 | 3479.0 | 3927.6 | 8.7012 |
| 800 | 2.47550 | 3664.7 | 4159.8 | 9.2479 | 1.65004 | 3664.3 | 4159.3 | 9.0605 | 1.23730 | 3663.9 | 4158.9 | 8.9274 |
| 900 | 2.70656 | 3856.3 | 4397.7 | 9.4598 | 1.80417 | 3856.0 | 4397.3 | 9.2725 | 1.35298 | 3855.7 | 4396.9 | 9.1394 |
| 1000 | 2.93755 | 4054.8 | 4642.3 | 9.6599 | 1.95824 | 4054.5 | 4642.0 | 9.4726 | 1.46859 | 4054.3 | 4641.7 | 9.3396 |
| 1100 | 3.16848 | 4259.6 | 4893.3 | 9.8497 | 2.11226 | 4259.4 | 4893.1 | 9.6624 | 1.58414 | 4259.2 | 4892.9 | 9.5295 |
| 1200 | 3.39938 | 4470.5 | 5150.4 | 10.0304 | 2.26624 | 4470.3 | 5150.2 | 9.8431 | 1.69966 | 4470.2 | 5150.0 | 9.7102 |
| 1300 | 3.63026 | 4687.1 | 5413.1 | 10.2029 | 2.42019 | 4686.9 | 5413.0 | 10.0157 | 1.81516 | 4686.7 | 5412.8 | 9.8828 |
| | $P = 0.50$ MPa (151.83°C) | | | | $P = 0.60$ MPa (158.83°C) | | | | $P = 0.80$ MPa (170.41°C) | | | |
| Sat. | 0.37483 | 2560.7 | 2748.1 | 6.8207 | 0.31560 | 2566.8 | 2756.2 | 6.7593 | 0.24035 | 2576.0 | 2768.3 | 6.6616 |
| 200 | 0.42503 | 2643.3 | 2855.8 | 7.0610 | 0.35212 | 2639.4 | 2850.6 | 6.9683 | 0.26088 | 2631.1 | 2839.8 | 6.8177 |
| 250 | 0.47443 | 2723.8 | 2961.0 | 7.2725 | 0.39390 | 2721.2 | 2957.6 | 7.1833 | 0.29321 | 2715.9 | 2950.4 | 7.0402 |
| 300 | 0.52261 | 2803.3 | 3064.6 | 7.4614 | 0.43442 | 2801.4 | 3062.0 | 7.3740 | 0.32416 | 2797.5 | 3056.9 | 7.2345 |
| 350 | 0.57015 | 2883.0 | 3168.1 | 7.6346 | 0.47428 | 2881.6 | 3166.1 | 7.5481 | 0.35442 | 2878.6 | 3162.2 | 7.4107 |
| 400 | 0.61731 | 2963.7 | 3272.4 | 7.7956 | 0.51374 | 2962.5 | 3270.8 | 7.7097 | 0.38429 | 2960.2 | 3267.7 | 7.5735 |
| 500 | 0.71095 | 3129.0 | 3484.5 | 8.0893 | 0.59200 | 3128.2 | 3483.4 | 8.0041 | 0.44332 | 3126.6 | 3481.3 | 7.8692 |
| 600 | 0.80409 | 3300.4 | 3702.5 | 8.3544 | 0.66976 | 3299.8 | 3701.7 | 8.2695 | 0.50186 | 3298.7 | 3700.1 | 8.1354 |
| 700 | 0.89696 | 3478.6 | 3927.0 | 8.5978 | 0.74725 | 3478.1 | 3926.4 | 8.5132 | 0.56011 | 3477.2 | 3925.3 | 8.3794 |
| 800 | 0.98966 | 3663.6 | 4158.4 | 8.8240 | 0.82457 | 3663.2 | 4157.9 | 8.7395 | 0.61820 | 3662.5 | 4157.0 | 8.6061 |
| 900 | 1.08227 | 3855.4 | 4396.6 | 9.0362 | 0.90179 | 3855.1 | 4396.2 | 8.9518 | 0.67619 | 3854.5 | 4395.5 | 8.8185 |
| 1000 | 1.17480 | 4054.0 | 4641.4 | 9.2364 | 0.97893 | 4053.8 | 4641.1 | 9.1521 | 0.73411 | 4053.3 | 4640.5 | 9.0189 |
| 1100 | 1.26728 | 4259.0 | 4892.6 | 9.4263 | 1.05603 | 4258.8 | 4892.4 | 9.3420 | 0.79197 | 4258.3 | 4891.9 | 9.2090 |
| 1200 | 1.35972 | 4470.0 | 5149.8 | 9.6071 | 1.13309 | 4469.8 | 5149.6 | 9.5229 | 0.84980 | 4469.4 | 5149.3 | 9.3898 |
| 1300 | 1.45214 | 4686.6 | 5412.6 | 9.7797 | 1.21012 | 4686.4 | 5412.5 | 9.6955 | 0.90761 | 4686.1 | 5412.2 | 9.5625 |

*The temperature in parentheses is the saturation temperature at the specified pressure.

†Properties of saturated vapor at the specified pressure.

Superheated water *(Continued)*

| T °C | v m³/kg | u kJ/kg | h kJ/kg | s kJ/kg·K | v m³/kg | u kJ/kg | h kJ/kg | s kJ/kg·K | v m³/kg | u kJ/kg | h kJ/kg | s kJ/kg·K |
|------|---------|---------|---------|-----------|---------|---------|---------|-----------|---------|---------|---------|-----------|
| | *P* = 1.00 MPa (179.88°C) | | | | *P* = 1.20 MPa (187.96°C) | | | | *P* = 1.40 MPa (195.04°C) | | | |
| Sat. | 0.19437 | 2582.8 | 2777.1 | 6.5850 | 0.16326 | 2587.8 | 2783.8 | 6.5217 | 0.14078 | 2591.8 | 2788.9 | 6.4675 |
| 200 | 0.20602 | 2622.3 | 2828.3 | 6.6956 | 0.16934 | 2612.9 | 2816.1 | 6.5909 | 0.14303 | 2602.7 | 2803.0 | 6.4975 |
| 250 | 0.23275 | 2710.4 | 2943.1 | 6.9265 | 0.19241 | 2704.7 | 2935.6 | 6.8313 | 0.16356 | 2698.9 | 2927.9 | 6.7488 |
| 300 | 0.25799 | 2793.7 | 3051.6 | 7.1246 | 0.21386 | 2789.7 | 3046.3 | 7.0335 | 0.18233 | 2785.7 | 3040.9 | 6.9553 |
| 350 | 0.28250 | 2875.7 | 3158.2 | 7.3029 | 0.23455 | 2872.7 | 3154.2 | 7.2139 | 0.20029 | 2869.7 | 3150.1 | 7.1379 |
| 400 | 0.30661 | 2957.9 | 3264.5 | 7.4670 | 0.25482 | 2955.5 | 3261.3 | 7.3793 | 0.21782 | 2953.1 | 3258.1 | 7.3046 |
| 500 | 0.35411 | 3125.0 | 3479.1 | 7.7642 | 0.29464 | 3123.4 | 3477.0 | 7.6779 | 0.25216 | 3121.8 | 3474.8 | 7.6047 |
| 600 | 0.40111 | 3297.5 | 3698.6 | 8.0311 | 0.33395 | 3296.3 | 3697.0 | 7.9456 | 0.28597 | 3295.1 | 3695.5 | 7.8730 |
| 700 | 0.44783 | 3476.3 | 3924.1 | 8.2755 | 0.37297 | 3475.3 | 3922.9 | 8.1904 | 0.31951 | 3474.4 | 3921.7 | 8.1183 |
| 800 | 0.49438 | 3661.7 | 4156.1 | 8.5024 | 0.41184 | 3661.0 | 4155.2 | 8.4176 | 0.35288 | 3660.3 | 4154.3 | 8.3458 |
| 900 | 0.54083 | 3853.9 | 4394.8 | 8.7150 | 0.45059 | 3853.3 | 4394.0 | 8.6303 | 0.38614 | 3852.7 | 4393.3 | 8.5587 |
| 1000 | 0.58721 | 4052.7 | 4640.0 | 8.9155 | 0.48928 | 4052.2 | 4639.4 | 8.8310 | 0.41933 | 4051.7 | 4638.8 | 8.7595 |
| 1100 | 0.63354 | 4257.9 | 4891.4 | 9.1057 | 0.52792 | 4257.5 | 4891.0 | 9.0212 | 0.45247 | 4257.0 | 4890.5 | 8.9497 |
| 1200 | 0.67983 | 4469.0 | 5148.9 | 9.2866 | 0.56652 | 4468.7 | 5148.5 | 9.2022 | 0.48558 | 4468.3 | 5148.1 | 9.1308 |
| 1300 | 0.72610 | 4685.8 | 5411.9 | 9.4593 | 0.60509 | 4685.5 | 5411.6 | 9.3750 | 0.51866 | 4685.1 | 5411.3 | 9.3036 |
| | *P* = 1.60 MPa (201.37°C) | | | | *P* = 1.80 MPa (207.11°C) | | | | *P* = 2.00 MPa (212.38°C) | | | |
| Sat. | 0.12374 | 2594.8 | 2792.8 | 6.4200 | 0.11037 | 2597.3 | 2795.9 | 6.3775 | 0.09959 | 2599.1 | 2798.3 | 6.3390 |
| 225 | 0.13293 | 2645.1 | 2857.8 | 6.5537 | 0.11678 | 2637.0 | 2847.2 | 6.4825 | 0.10381 | 2628.5 | 2836.1 | 6.4160 |
| 250 | 0.14190 | 2692.9 | 2919.9 | 6.6753 | 0.12502 | 2686.7 | 2911.7 | 6.6088 | 0.11150 | 2680.3 | 2903.3 | 6.5475 |
| 300 | 0.15866 | 2781.6 | 3035.4 | 6.8864 | 0.14025 | 2777.4 | 3029.9 | 6.8246 | 0.12551 | 2773.2 | 3024.2 | 6.7684 |
| 350 | 0.17459 | 2866.6 | 3146.0 | 7.0713 | 0.15460 | 2863.6 | 3141.9 | 7.0120 | 0.13860 | 2860.5 | 3137.7 | 6.9583 |
| 400 | 0.19007 | 2950.8 | 3254.9 | 7.2394 | 0.16849 | 2948.3 | 3251.6 | 7.1814 | 0.15122 | 2945.9 | 3248.4 | 7.1292 |
| 500 | 0.22029 | 3120.1 | 3472.6 | 7.5410 | 0.19551 | 3118.5 | 3470.4 | 7.4845 | 0.17568 | 3116.9 | 3468.3 | 7.4337 |
| 600 | 0.24999 | 3293.9 | 3693.9 | 7.8101 | 0.22200 | 3292.7 | 3692.3 | 7.7543 | 0.19962 | 3291.5 | 3690.7 | 7.7043 |
| 700 | 0.27941 | 3473.5 | 3920.5 | 8.0558 | 0.24822 | 3472.6 | 3919.4 | 8.0005 | 0.22326 | 3471.7 | 3918.2 | 7.9509 |
| 800 | 0.30865 | 3659.5 | 4153.4 | 8.2834 | 0.27426 | 3658.8 | 4152.4 | 8.2284 | 0.24674 | 3658.0 | 4151.5 | 8.1791 |
| 900 | 0.33780 | 3852.1 | 4392.6 | 8.4965 | 0.30020 | 3851.5 | 4391.9 | 8.4417 | 0.27012 | 3850.9 | 4391.1 | 8.3925 |
| 1000 | 0.36687 | 4051.2 | 4638.2 | 8.6974 | 0.32606 | 4050.7 | 4637.6 | 8.6427 | 0.29342 | 4050.2 | 4637.1 | 8.5936 |
| 1100 | 0.39589 | 4256.6 | 4890.0 | 8.8878 | 0.35188 | 4256.2 | 4889.6 | 8.8331 | 0.31667 | 4255.7 | 4889.1 | 8.7842 |
| 1200 | 0.42488 | 4467.9 | 5147.7 | 9.0689 | 0.37766 | 4467.6 | 5147.3 | 9.0143 | 0.33989 | 4467.2 | 5147.0 | 8.9654 |
| 1300 | 0.45383 | 4684.8 | 5410.9 | 9.2418 | 0.40341 | 4684.5 | 5410.6 | 9.1872 | 0.36308 | 4684.2 | 5410.3 | 9.1384 |
| | *P* = 2.50 MPa (223.95°C) | | | | *P* = 3.00 MPa (233.85°C) | | | | *P* = 3.50 MPa (242.56°C) | | | |
| Sat. | 0.07995 | 2602.1 | 2801.9 | 6.2558 | 0.06667 | 2603.2 | 2803.2 | 6.1856 | 0.05706 | 2603.0 | 2802.7 | 6.1244 |
| 225 | 0.08026 | 2604.8 | 2805.5 | 6.2629 | | | | | | | | |
| 250 | 0.08705 | 2663.3 | 2880.9 | 6.4107 | 0.07063 | 2644.7 | 2856.5 | 6.2893 | 0.05876 | 2624.0 | 2829.7 | 6.1764 |
| 300 | 0.09894 | 2762.2 | 3009.6 | 26.6459 | 0.08118 | 2750.8 | 2994.3 | 6.5412 | 0.06845 | 2738.8 | 2978.4 | 6.4484 |
| 350 | 0.10979 | 2852.5 | 3127.0 | 6.8424 | 0.09056 | 2844.4 | 3116.1 | 6.7450 | 0.07680 | 2836.0 | 3104.9 | 6.6601 |
| 400 | 0.12012 | 2939.8 | 3240.1 | 7.0170 | 0.09938 | 2933.6 | 3231.7 | 6.9235 | 0.08456 | 2927.2 | 3223.2 | 6.8428 |
| 450 | 0.13015 | 3026.2 | 3351.6 | 7.1768 | 0.10789 | 3021.2 | 3344.9 | 7.0856 | 0.09198 | 3016.1 | 3338.1 | 7.0074 |
| 500 | 0.13999 | 3112.8 | 3462.8 | 7.3254 | 0.11620 | 3108.6 | 3457.2 | 7.2359 | 0.09919 | 3104.5 | 3451.7 | 7.1593 |
| 600 | 0.15931 | 3288.5 | 3686.8 | 7.5979 | 0.13245 | 3285.5 | 3682.8 | 7.5103 | 0.11325 | 3282.5 | 3678.9 | 7.4357 |
| 700 | 0.17835 | 3469.3 | 3915.2 | 7.8455 | 0.14841 | 3467.0 | 3912.2 | 7.7590 | 0.12702 | 3464.7 | 3909.3 | 7.6855 |
| 800 | 0.19722 | 3656.2 | 4149.2 | 8.0744 | 0.16420 | 3654.3 | 4146.9 | 7.9885 | 0.14061 | 3652.5 | 4144.6 | 7.9156 |
| 900 | 0.21597 | 3849.4 | 4389.3 | 8.2882 | 0.17988 | 3847.9 | 4387.5 | 8.2028 | 0.15410 | 3846.4 | 4385.7 | 8.1304 |
| 1000 | 0.23466 | 4049.0 | 4635.6 | 8.4897 | 0.19549 | 4047.7 | 4634.2 | 8.4045 | 0.16751 | 4046.4 | 4632.7 | 8.3324 |
| 1100 | 0.25330 | 4254.7 | 4887.9 | 8.6804 | 0.21105 | 4253.6 | 4886.7 | 8.5955 | 0.18087 | 4252.5 | 4885.6 | 8.5236 |
| 1200 | 0.27190 | 4466.3 | 5146.0 | 8.8618 | 0.22658 | 4465.3 | 5145.1 | 8.7771 | 0.19420 | 4464.4 | 5144.1 | 8.7053 |
| 1300 | 0.29048 | 4683.4 | 5409.5 | 9.0349 | 0.24207 | 4682.6 | 5408.8 | 8.9502 | 0.20750 | 4681.8 | 5408.0 | 8.8786 |

Superheated water *(Continued)*

| T °C | υ m³/kg | u kJ/kg | h kJ/kg | s kJ/kg·K | υ m³/kg | u kJ/kg | h kJ/kg | s kJ/kg·K | υ m³/kg | u kJ/kg | h kJ/kg | s kJ/kg·K |
|---|---|---|---|---|---|---|---|---|---|---|---|---|
| | *P* = 4.0 MPa (250.35°C) | | | | *P* = 4.5 MPa (257.44°C) | | | | *P* = 5.0 MPa (263.94°C) | | | |
| Sat. | 0.04978 | 2601.7 | 2800.8 | 6.0696 | 0.04406 | 2599.7 | 2798.0 | 6.0198 | 0.03945 | 2597.0 | 2794.2 | 5.9737 |
| 275 | 0.05461 | 2668.9 | 2887.3 | 6.2312 | 0.04733 | 2651.4 | 2864.4 | 6.1429 | 0.04144 | 2632.3 | 2839.5 | 6.0571 |
| 300 | 0.05887 | 2726.2 | 2961.7 | 6.3639 | 0.05138 | 2713.0 | 2944.2 | 6.2854 | 0.04535 | 2699.0 | 2925.7 | 6.2111 |
| 350 | 0.06647 | 2827.4 | 3093.3 | 6.5843 | 0.05842 | 2818.6 | 3081.5 | 6.5153 | 0.05197 | 2809.5 | 3069.3 | 6.4516 |
| 400 | 0.07343 | 2920.8 | 3214.5 | 6.7714 | 0.06477 | 2914.2 | 3205.7 | 6.7071 | 0.05784 | 2907.5 | 3196.7 | 6.6483 |
| 450 | 0.08004 | 3011.0 | 3331.2 | 6.9386 | 0.07076 | 3005.8 | 3324.2 | 6.8770 | 0.06332 | 3000.6 | 3317.2 | 6.8210 |
| 500 | 0.08644 | 3100.3 | 3446.0 | 7.0922 | 0.07652 | 3096.0 | 3440.4 | 7.0323 | 0.06858 | 3091.8 | 3434.7 | 6.9781 |
| 600 | 0.09886 | 3279.4 | 3674.9 | 7.3706 | 0.08766 | 3276.4 | 3670.9 | 7.3127 | 0.07870 | 3273.3 | 3666.9 | 7.2605 |
| 700 | 0.11098 | 3462.4 | 3906.3 | 7.6214 | 0.09850 | 3460.0 | 3903.3 | 7.5647 | 0.08852 | 3457.7 | 3900.3 | 7.5136 |
| 800 | 0.12292 | 3650.6 | 4142.3 | 7.8523 | 0.10916 | 3648.8 | 4140.0 | 7.7962 | 0.09816 | 3646.9 | 4137.7 | 7.7458 |
| 900 | 0.13476 | 3844.8 | 4383.9 | 8.0675 | 0.11972 | 3843.3 | 4382.1 | 8.0118 | 0.10769 | 3841.8 | 4380.2 | 7.9619 |
| 1000 | 0.14653 | 4045.1 | 4631.2 | 8.2698 | 0.13020 | 4043.9 | 4629.8 | 8.2144 | 0.11715 | 4042.6 | 4628.3 | 8.1648 |
| 1100 | 0.15824 | 4251.4 | 4884.4 | 8.4612 | 0.14064 | 4250.4 | 4883.2 | 8.4060 | 0.12655 | 4249.3 | 4882.1 | 8.3566 |
| 1200 | 0.16992 | 4463.5 | 5143.2 | 8.6430 | 0.15103 | 4462.6 | 5142.2 | 8.5880 | 0.13592 | 4461.6 | 5141.3 | 8.5388 |
| 1300 | 0.18157 | 4680.9 | 5407.2 | 8.8164 | 0.16140 | 4680.1 | 5406.5 | 8.7616 | 0.14527 | 4679.3 | 5405.7 | 8.7124 |
| | *P* = 6.0 MPa (275.59°C) | | | | *P* = 7.0 MPa (285.83°C) | | | | *P* = 8.0 MPa (295.01°C) | | | |
| Sat. | 0.03245 | 2589.9 | 2784.6 | 5.8902 | 0.027378 | 2581.0 | 2772.6 | 5.8148 | 0.023525 | 2570.5 | 2758.7 | 5.7450 |
| 300 | 0.03619 | 2668.4 | 2885.6 | 6.0703 | 0.029492 | 2633.5 | 2839.9 | 5.9337 | 0.024279 | 2592.3 | 2786.5 | 5.7937 |
| 350 | 0.04225 | 2790.4 | 3043.9 | 6.3357 | 0.035262 | 2770.1 | 3016.9 | 6.2305 | 0.029975 | 2748.3 | 2988.1 | 6.1321 |
| 400 | 0.04742 | 2893.7 | 3178.3 | 6.5432 | 0.039958 | 2879.5 | 3159.2 | 6.4502 | 0.034344 | 2864.6 | 3139.4 | 6.3658 |
| 450 | 0.05217 | 2989.9 | 3302.9 | 6.7219 | 0.044187 | 2979.0 | 3288.3 | 6.6353 | 0.038194 | 2967.8 | 3273.3 | 6.5579 |
| 500 | 0.05667 | 3083.1 | 3423.1 | 6.8826 | 0.048157 | 3074.3 | 3411.4 | 6.8000 | 0.041767 | 3065.4 | 3399.5 | 6.7266 |
| 550 | 0.06102 | 3175.2 | 3541.3 | 7.0308 | 0.051966 | 3167.9 | 3531.6 | 6.9507 | 0.045172 | 3160.5 | 3521.8 | 6.8800 |
| 600 | 0.06527 | 3267.2 | 3658.8 | 7.1693 | 0.055665 | 3261.0 | 3650.6 | 7.0910 | 0.048463 | 3254.7 | 3642.4 | 7.0221 |
| 700 | 0.07355 | 3453.0 | 3894.3 | 7.4247 | 0.062850 | 3448.3 | 3888.3 | 7.3487 | 0.054829 | 3443.6 | 3882.2 | 7.2822 |
| 800 | 0.08165 | 3643.2 | 4133.1 | 7.6582 | 0.069856 | 3639.5 | 4128.5 | 7.5836 | 0.061011 | 3635.7 | 4123.8 | 7.5185 |
| 900 | 0.08964 | 3838.8 | 4376.6 | 7.8751 | 0.076750 | 3835.7 | 4373.0 | 7.8014 | 0.067082 | 3832.7 | 4369.3 | 7.7372 |
| 1000 | 0.09756 | 4040.1 | 4625.4 | 8.0786 | 0.083571 | 4037.5 | 4622.5 | 8.0055 | 0.073079 | 4035.0 | 4619.6 | 7.9419 |
| 1100 | 0.10543 | 4247.1 | 4879.7 | 8.2709 | 0.090341 | 4245.0 | 4877.4 | 8.1982 | 0.079025 | 4242.8 | 4875.0 | 8.1350 |
| 1200 | 0.11326 | 4459.8 | 5139.4 | 8.4534 | 0.097075 | 4457.9 | 5137.4 | 8.3810 | 0.084934 | 4456.1 | 5135.5 | 8.3181 |
| 1300 | 0.12107 | 4677.7 | 5404.1 | 8.6273 | 0.103781 | 4676.1 | 5402.6 | 8.5551 | 0.090817 | 4674.5 | 5401.0 | 8.4925 |
| | *P* = 9.0 MPa (303.35°C) | | | | *P* = 10.0 MPa (311.00°C) | | | | *P* = 12.5 MPa (327.81°C) | | | |
| Sat. | 0.020489 | 2558.5 | 2742.9 | 5.6791 | 0.018028 | 2545.2 | 2725.5 | 5.6159 | 0.013496 | 2505.6 | 2674.3 | 5.4638 |
| 325 | 0.023284 | 2647.6 | 2857.1 | 5.8738 | 0.019877 | 2611.6 | 2810.3 | 5.7596 | | | | |
| 350 | 0.025816 | 2725.0 | 2957.3 | 6.0380 | 0.022440 | 2699.6 | 2924.0 | 5.9460 | 0.016138 | 2624.9 | 2826.6 | 5.7130 |
| 400 | 0.029960 | 2849.2 | 3118.8 | 6.2876 | 0.026436 | 2833.1 | 3097.5 | 6.2141 | 0.020030 | 2789.6 | 3040.0 | 6.0433 |
| 450 | 0.033524 | 2956.3 | 3258.0 | 6.4872 | 0.029782 | 2944.5 | 3242.4 | 6.4219 | 0.023019 | 2913.7 | 3201.5 | 6.2749 |
| 500 | 0.036793 | 3056.3 | 3387.4 | 6.6603 | 0.032811 | 3047.0 | 3375.1 | 6.5995 | 0.025630 | 3023.2 | 3343.6 | 6.4651 |
| 550 | 0.039885 | 3153.0 | 3512.0 | 6.8164 | 0.035655 | 3145.4 | 3502.0 | 6.7585 | 0.028033 | 3126.1 | 3476.5 | 6.6317 |
| 600 | 0.042861 | 3248.4 | 3634.1 | 6.9605 | 0.038378 | 3242.0 | 3625.8 | 6.9045 | 0.030306 | 3225.8 | 3604.6 | 6.7828 |
| 650 | 0.045755 | 3343.4 | 3755.2 | 7.0954 | 0.041018 | 3338.0 | 3748.1 | 7.0408 | 0.032491 | 3324.1 | 3730.2 | 6.9227 |
| 700 | 0.048589 | 3438.8 | 3876.1 | 7.2229 | 0.043597 | 3434.0 | 3870.0 | 7.1693 | 0.034612 | 3422.0 | 3854.6 | 7.0540 |
| 800 | 0.054132 | 3632.0 | 4119.2 | 7.4606 | 0.048629 | 3628.2 | 4114.5 | 7.4085 | 0.038724 | 3618.8 | 4102.8 | 7.2967 |
| 900 | 0.059562 | 3829.6 | 4365.7 | 7.6802 | 0.053547 | 3826.5 | 4362.0 | 7.6290 | 0.042720 | 3818.9 | 4352.9 | 7.5195 |
| 1000 | 0.064919 | 4032.4 | 4616.7 | 7.8855 | 0.058391 | 4029.9 | 4613.8 | 7.8349 | 0.046641 | 4023.5 | 4606.5 | 7.7269 |
| 1100 | 0.070224 | 4240.7 | 4872.7 | 8.0791 | 0.063183 | 4238.5 | 4870.3 | 8.0289 | 0.050510 | 4233.1 | 4864.5 | 7.9220 |
| 1200 | 0.075492 | 4454.2 | 5133.6 | 8.2625 | 0.067938 | 4452.4 | 5131.7 | 8.2126 | 0.054342 | 4447.7 | 5127.0 | 8.1065 |
| 1300 | 0.080733 | 4672.9 | 5399.5 | 8.4371 | 0.072667 | 4671.3 | 5398.0 | 8.3874 | 0.058147 | 4667.3 | 5394.1 | 8.2819 |

14

Superheated water *(Concluded)*

| T °C | υ m³/kg | u kJ/kg | h kJ/kg | s kJ/kg·K | υ m³/kg | u kJ/kg | h kJ/kg | s kJ/kg·K | υ m³/kg | u kJ/kg | h kJ/kg | s kJ/kg·K |
|---|---|---|---|---|---|---|---|---|---|---|---|---|
| | *P* = 15.0 MPa (342.16°C) | | | | *P* = 17.5 MPa (354.67°C) | | | | *P* = 20.0 MPa (365.75°C) | | | |
| Sat. | 0.010341 | 2455.7 | 2610.8 | 5.3108 | 0.007932 | 2390.7 | 2529.5 | 5.1435 | 0.005862 | 2294.8 | 2412.1 | 4.9310 |
| 350 | 0.011481 | 2520.9 | 2693.1 | 5.4438 | | | | | | | | |
| 400 | 0.015671 | 2740.6 | 2975.7 | 5.8819 | 0.012463 | 2684.3 | 2902.4 | 5.7211 | 0.009950 | 2617.9 | 2816.9 | 5.5526 |
| 450 | 0.018477 | 2880.8 | 3157.9 | 6.1434 | 0.015204 | 2845.4 | 3111.4 | 6.0212 | 0.012721 | 2807.3 | 3061.7 | 5.9043 |
| 500 | 0.020828 | 2998.4 | 3310.8 | 6.3480 | 0.017385 | 2972.4 | 3276.7 | 6.2424 | 0.014793 | 2945.3 | 3241.2 | 6.1446 |
| 550 | 0.022945 | 3106.2 | 3450.4 | 6.5230 | 0.019305 | 3085.8 | 3423.6 | 6.4266 | 0.016571 | 3064.7 | 3396.2 | 6.3390 |
| 600 | 0.024921 | 3209.3 | 3583.1 | 6.6796 | 0.021073 | 3192.5 | 3561.3 | 6.5890 | 0.018185 | 3175.3 | 3539.0 | 6.5075 |
| 650 | 0.026804 | 3310.1 | 3712.1 | 6.8233 | 0.022742 | 3295.8 | 3693.8 | 6.7366 | 0.019695 | 3281.4 | 3675.3 | 6.6593 |
| 700 | 0.028621 | 3409.8 | 3839.1 | 6.9573 | 0.024342 | 3397.5 | 3823.5 | 6.8735 | 0.021134 | 3385.1 | 3807.8 | 6.7991 |
| 800 | 0.032121 | 3609.3 | 4091.1 | 7.2037 | 0.027405 | 3599.7 | 4079.3 | 7.1237 | 0.023870 | 3590.1 | 4067.5 | 7.0531 |
| 900 | 0.035503 | 3811.2 | 4343.7 | 7.4288 | 0.030348 | 3803.5 | 4334.6 | 7.3511 | 0.026484 | 3795.7 | 4325.4 | 7.2829 |
| 1000 | 0.038808 | 4017.1 | 4599.2 | 7.6378 | 0.033215 | 4010.7 | 4592.0 | 7.5616 | 0.029020 | 4004.3 | 4584.7 | 7.4950 |
| 1100 | 0.042062 | 4227.7 | 4858.6 | 7.8339 | 0.036029 | 4222.3 | 4852.8 | 7.7588 | 0.031504 | 4216.9 | 4847.0 | 7.6933 |
| 1200 | 0.045279 | 4443.1 | 5122.3 | 8.0192 | 0.038806 | 4438.5 | 5117.6 | 7.9449 | 0.033952 | 4433.8 | 5112.9 | 7.8802 |
| 1300 | 0.048469 | 4663.3 | 5390.3 | 8.1952 | 0.041556 | 4659.2 | 5386.5 | 8.1215 | 0.036371 | 4655.2 | 5382.7 | 8.0574 |
| | *P* = 25.0 MPa | | | | *P* = 30.0 MPa | | | | *P* = 35.0 MPa | | | |
| 375 | 0.001978 | 1799.9 | 1849.4 | 4.0345 | 0.001792 | 1738.1 | 1791.9 | 3.9313 | 0.001701 | 1702.8 | 1762.4 | 3.8724 |
| 400 | 0.006005 | 2428.5 | 2578.7 | 5.1400 | 0.002798 | 2068.9 | 2152.8 | 4.4758 | 0.002105 | 1914.9 | 1988.6 | 4.2144 |
| 425 | 0.007886 | 2607.8 | 2805.0 | 5.4708 | 0.005299 | 2452.9 | 2611.8 | 5.1473 | 0.003434 | 2253.3 | 2373.5 | 4.7751 |
| 450 | 0.009176 | 2721.2 | 2950.6 | 5.6759 | 0.006737 | 2618.9 | 2821.0 | 5.4422 | 0.004957 | 2497.5 | 2671.0 | 5.1946 |
| 500 | 0.011143 | 2887.3 | 3165.9 | 5.9643 | 0.008691 | 2824.0 | 3084.8 | 5.7956 | 0.006933 | 2755.3 | 2997.9 | 5.6331 |
| 550 | 0.012736 | 3020.8 | 3339.2 | 6.1816 | 0.010175 | 2974.5 | 3279.7 | 6.0403 | 0.008348 | 2925.8 | 3218.0 | 5.9093 |
| 600 | 0.014140 | 3140.0 | 3493.5 | 6.3637 | 0.011445 | 3103.4 | 3446.8 | 6.2373 | 0.009523 | 3065.6 | 3399.0 | 6.1229 |
| 650 | 0.015430 | 3251.9 | 3637.7 | 6.5243 | 0.012590 | 3221.7 | 3599.4 | 6.4074 | 0.010565 | 3190.9 | 3560.7 | 6.3030 |
| 700 | 0.016643 | 3359.9 | 3776.0 | 6.6702 | 0.013654 | 3334.3 | 3743.9 | 6.5599 | 0.011523 | 3308.3 | 3711.6 | 6.4623 |
| 800 | 0.018922 | 3570.7 | 4043.8 | 6.9322 | 0.015628 | 3551.2 | 4020.0 | 6.8301 | 0.013278 | 3531.6 | 3996.3 | 6.7409 |
| 900 | 0.021075 | 3780.2 | 4307.1 | 7.1668 | 0.017473 | 3764.6 | 4288.8 | 7.0695 | 0.014904 | 3749.0 | 4270.6 | 6.9853 |
| 1000 | 0.023150 | 3991.5 | 4570.2 | 7.3821 | 0.019240 | 3978.6 | 4555.8 | 7.2880 | 0.016450 | 3965.8 | 4541.5 | 7.2069 |
| 1100 | 0.025172 | 4206.1 | 4835.4 | 7.5825 | 0.020954 | 4195.2 | 4823.9 | 7.4906 | 0.017942 | 4184.4 | 4812.4 | 7.4118 |
| 1200 | 0.027157 | 4424.6 | 5103.5 | 7.7710 | 0.022630 | 4415.3 | 5094.2 | 7.6807 | 0.019398 | 4406.1 | 5085.0 | 7.6034 |
| 1300 | 0.029115 | 4647.2 | 5375.1 | 7.9494 | 0.024279 | 4639.2 | 5367.6 | 7.8602 | 0.020827 | 4631.2 | 5360.2 | 7.7841 |
| | *P* = 40.0 MPa | | | | *P* = 50.0 MPa | | | | *P* = 60.0 MPa | | | |
| 375 | 0.001641 | 1677.0 | 1742.6 | 3.8290 | 0.001560 | 1638.6 | 1716.6 | 3.7642 | 0.001503 | 1609.7 | 1699.9 | 3.7149 |
| 400 | 0.001911 | 1855.0 | 1931.4 | 4.1145 | 0.001731 | 1787.8 | 1874.4 | 4.0029 | 0.001633 | 1745.2 | 1843.2 | 3.9317 |
| 425 | 0.002538 | 2097.5 | 2199.0 | 4.5044 | 0.002009 | 1960.3 | 2060.7 | 4.2746 | 0.001816 | 1892.9 | 2001.8 | 4.1630 |
| 450 | 0.003692 | 2364.2 | 2511.8 | 4.9449 | 0.002487 | 2160.3 | 2284.7 | 4.5896 | 0.002086 | 2055.1 | 2180.2 | 4.4140 |
| 500 | 0.005623 | 2681.6 | 2906.5 | 5.4744 | 0.003890 | 2528.1 | 2722.6 | 5.1762 | 0.002952 | 2393.2 | 2570.3 | 4.9356 |
| 550 | 0.006985 | 2875.1 | 3154.4 | 5.7857 | 0.005118 | 2769.5 | 3025.4 | 5.5563 | 0.003955 | 2664.6 | 2901.9 | 5.3517 |
| 600 | 0.008089 | 3026.8 | 3350.4 | 6.0170 | 0.006108 | 2947.1 | 3252.6 | 5.8245 | 0.004833 | 2866.8 | 3156.8 | 5.6527 |
| 650 | 0.009053 | 3159.5 | 3521.6 | 6.2078 | 0.006957 | 3095.6 | 3443.5 | 6.0373 | 0.005591 | 3031.3 | 3366.8 | 5.8867 |
| 700 | 0.009930 | 3282.0 | 3679.2 | 6.3740 | 0.007717 | 3228.7 | 3614.6 | 6.2179 | 0.006265 | 3175.4 | 3551.3 | 6.0814 |
| 800 | 0.011521 | 3511.8 | 3972.6 | 6.6613 | 0.009073 | 3472.2 | 3925.8 | 6.5225 | 0.007456 | 3432.6 | 3880.0 | 6.4033 |
| 900 | 0.012980 | 3733.3 | 4252.5 | 6.9107 | 0.010296 | 3702.0 | 4216.8 | 6.7819 | 0.008519 | 3670.9 | 4182.1 | 6.6725 |
| 1000 | 0.014360 | 3952.9 | 4527.3 | 7.1355 | 0.011441 | 3927.4 | 4499.4 | 7.0131 | 0.009504 | 3902.0 | 4472.2 | 6.9099 |
| 1100 | 0.015686 | 4173.7 | 4801.1 | 7.3425 | 0.012534 | 4152.2 | 4778.9 | 7.2244 | 0.010439 | 4130.9 | 4757.3 | 7.1255 |
| 1200 | 0.016976 | 4396.9 | 5075.9 | 7.5357 | 0.013590 | 4378.6 | 5058.1 | 7.4207 | 0.011339 | 4360.5 | 5040.8 | 7.3248 |
| 1300 | 0.018239 | 4623.3 | 5352.8 | 7.7175 | 0.014620 | 4607.5 | 5338.5 | 7.6048 | 0.012213 | 4591.8 | 5324.5 | 7.5111 |

## TABLE A–7

Compressed liquid water

| T °C | υ m³/kg | u kJ/kg | h kJ/kg | s kJ/kg·K | υ m³/kg | u kJ/kg | h kJ/kg | s kJ/kg·K | υ m³/kg | u kJ/kg | h kJ/kg | s kJ/kg·K |
|---|---|---|---|---|---|---|---|---|---|---|---|---|
| | *P* = 5 MPa (263.94°C) | | | | *P* = 10 MPa (311.00°C) | | | | *P* = 15 MPa (342.16°C) | | | |
| Sat. | 0.0012862 | 1148.1 | 1154.5 | 2.9207 | 0.0014522 | 1393.3 | 1407.9 | 3.3603 | 0.0016572 | 1585.5 | 1610.3 | 3.6848 |
| 0 | 0.0009977 | 0.04 | 5.03 | 0.0001 | 0.0009952 | 0.12 | 10.07 | 0.0003 | 0.0009928 | 0.18 | 15.07 | 0.0004 |
| 20 | 0.0009996 | 83.61 | 88.61 | 0.2954 | 0.0009973 | 83.31 | 93.28 | 0.2943 | 0.0009951 | 83.01 | 97.93 | 0.2932 |
| 40 | 0.0010057 | 166.92 | 171.95 | 0.5705 | 0.0010035 | 166.33 | 176.37 | 0.5685 | 0.0010013 | 165.75 | 180.77 | 0.5666 |
| 60 | 0.0010149 | 250.29 | 255.36 | 0.8287 | 0.0010127 | 249.43 | 259.55 | 0.8260 | 0.0010105 | 248.58 | 263.74 | 0.8234 |
| 80 | 0.0010267 | 333.82 | 338.96 | 1.0723 | 0.0010244 | 332.69 | 342.94 | 1.0691 | 0.0010221 | 331.59 | 346.92 | 1.0659 |
| 100 | 0.0010410 | 417.65 | 422.85 | 1.3034 | 0.0010385 | 416.23 | 426.62 | 1.2996 | 0.0010361 | 414.85 | 430.39 | 1.2958 |
| 120 | 0.0010576 | 501.91 | 507.19 | 1.5236 | 0.0010549 | 500.18 | 510.73 | 1.5191 | 0.0010522 | 498.50 | 514.28 | 1.5148 |
| 140 | 0.0010769 | 586.80 | 592.18 | 1.7344 | 0.0010738 | 584.72 | 595.45 | 1.7293 | 0.0010708 | 582.69 | 598.75 | 1.7243 |
| 160 | 0.0010988 | 672.55 | 678.04 | 1.9374 | 0.0010954 | 670.06 | 681.01 | 1.9316 | 0.0010920 | 667.63 | 684.01 | 1.9259 |
| 180 | 0.0011240 | 759.47 | 765.09 | 2.1338 | 0.0011200 | 756.48 | 767.68 | 2.1271 | 0.0011160 | 753.58 | 770.32 | 2.1206 |
| 200 | 0.0011531 | 847.92 | 853.68 | 2.3251 | 0.0011482 | 844.32 | 855.80 | 2.3174 | 0.0011435 | 840.84 | 858.00 | 2.3100 |
| 220 | 0.0011868 | 938.39 | 944.32 | 2.5127 | 0.0011809 | 934.01 | 945.82 | 2.5037 | 0.0011752 | 929.81 | 947.43 | 2.4951 |
| 240 | 0.0012268 | 1031.6 | 1037.7 | 2.6983 | 0.0012192 | 1026.2 | 1038.3 | 2.6876 | 0.0012121 | 1021.0 | 1039.2 | 2.6774 |
| 260 | 0.0012755 | 1128.5 | 1134.9 | 2.8841 | 0.0012653 | 1121.6 | 1134.3 | 2.8710 | 0.0012560 | 1115.1 | 1134.0 | 2.8586 |
| 280 | | | | | 0.0013226 | 1221.8 | 1235.0 | 3.0565 | 0.0013096 | 1213.4 | 1233.0 | 3.0410 |
| 300 | | | | | 0.0013980 | 1329.4 | 1343.3 | 3.2488 | 0.0013783 | 1317.6 | 1338.3 | 3.2279 |
| 320 | | | | | | | | | 0.0014733 | 1431.9 | 1454.0 | 3.4263 |
| 340 | | | | | | | | | 0.0016311 | 1567.9 | 1592.4 | 3.6555 |
| | *P* = 20 MPa (365.75°C) | | | | *P* = 30 MPa | | | | *P* = 50 MPa | | | |
| Sat. | 0.0020378 | 1785.8 | 1826.6 | 4.0146 | | | | | | | | |
| 0 | 0.0009904 | 0.23 | 20.03 | 0.0005 | 0.0009857 | 0.29 | 29.86 | 0.0003 | 0.0009767 | 0.29 | 49.13 | −0.0010 |
| 20 | 0.0009929 | 82.71 | 102.57 | 0.2921 | 0.0009886 | 82.11 | 111.77 | 0.2897 | 0.0009805 | 80.93 | 129.95 | 0.2845 |
| 40 | 0.0009992 | 165.17 | 185.16 | 0.5646 | 0.0009951 | 164.05 | 193.90 | 0.5607 | 0.0009872 | 161.90 | 211.25 | 0.5528 |
| 60 | 0.0010084 | 247.75 | 267.92 | 0.8208 | 0.0010042 | 246.14 | 276.26 | 0.8156 | 0.0009962 | 243.08 | 292.88 | 0.8055 |
| 80 | 0.0010199 | 330.50 | 350.90 | 1.0627 | 0.0010155 | 328.40 | 358.86 | 1.0564 | 0.0010072 | 324.42 | 374.78 | 1.0442 |
| 100 | 0.0010337 | 413.50 | 434.17 | 1.2920 | 0.0010290 | 410.87 | 441.74 | 1.2847 | 0.0010201 | 405.94 | 456.94 | 1.2705 |
| 120 | 0.0010496 | 496.85 | 517.84 | 1.5105 | 0.0010445 | 493.66 | 525.00 | 1.5020 | 0.0010349 | 487.69 | 539.43 | 1.4859 |
| 140 | 0.0010679 | 580.71 | 602.07 | 1.7194 | 0.0010623 | 576.90 | 608.76 | 1.7098 | 0.0010517 | 569.77 | 622.36 | 1.6916 |
| 160 | 0.0010886 | 665.28 | 687.05 | 1.9203 | 0.0010823 | 660.74 | 693.21 | 1.9094 | 0.0010704 | 652.33 | 705.85 | 1.8889 |
| 180 | 0.0011122 | 750.78 | 773.02 | 2.1143 | 0.0011049 | 745.40 | 778.55 | 2.1020 | 0.0010914 | 735.49 | 790.06 | 2.0790 |
| 200 | 0.0011390 | 837.49 | 860.27 | 2.3027 | 0.0011304 | 831.11 | 865.02 | 2.2888 | 0.0011149 | 819.45 | 875.19 | 2.2628 |
| 220 | 0.0011697 | 925.77 | 949.16 | 2.4867 | 0.0011595 | 918.15 | 952.93 | 2.4707 | 0.0011412 | 904.39 | 961.45 | 2.4414 |
| 240 | 0.0012053 | 1016.1 | 1040.2 | 2.6676 | 0.0011927 | 1006.9 | 1042.7 | 2.6491 | 0.0011708 | 990.55 | 1049.1 | 2.6156 |
| 260 | 0.0012472 | 1109.0 | 1134.0 | 2.8469 | 0.0012314 | 1097.8 | 1134.7 | 2.8250 | 0.0012044 | 1078.2 | 1138.4 | 2.7864 |
| 280 | 0.0012978 | 1205.6 | 1231.5 | 3.0265 | 0.0012770 | 1191.5 | 1229.8 | 3.0001 | 0.0012430 | 1167.7 | 1229.9 | 2.9547 |
| 300 | 0.0013611 | 1307.2 | 1334.4 | 3.2091 | 0.0013322 | 1288.9 | 1328.9 | 3.1761 | 0.0012879 | 1259.6 | 1324.0 | 3.1218 |
| 320 | 0.0014450 | 1416.6 | 1445.5 | 3.3996 | 0.0014014 | 1391.7 | 1433.7 | 3.3558 | 0.0013409 | 1354.3 | 1421.4 | 3.2888 |
| 340 | 0.0015693 | 1540.2 | 1571.6 | 3.6086 | 0.0014932 | 1502.4 | 1547.1 | 3.5438 | 0.0014049 | 1452.9 | 1523.1 | 3.4575 |
| 360 | 0.0018248 | 1703.6 | 1740.1 | 3.8787 | 0.0016276 | 1626.8 | 1675.6 | 3.7499 | 0.0014848 | 1556.5 | 1630.7 | 3.6301 |
| 380 | | | | | 0.0018729 | 1782.0 | 1838.2 | 4.0026 | 0.0015884 | 1667.1 | 1746.5 | 3.8102 |

## TABLE A–8

Saturated ice–water vapor

| Temp., $T$ °C | Sat. press., $P_{sat}$ kPa | Specific volume, m³/kg | | Internal energy, kJ/kg | | | Enthalpy, kJ/kg | | | Entropy, kJ/kg·K | | |
|---|---|---|---|---|---|---|---|---|---|---|---|---|
| | | Sat. ice, $v_i$ | Sat. vapor, $v_g$ | Sat. ice, $u_i$ | Subl., $u_{ig}$ | Sat. vapor, $u_g$ | Sat. ice, $h_i$ | Subl., $h_{ig}$ | Sat. vapor, $h_g$ | Sat. ice, $s_i$ | Subl., $s_{ig}$ | Sat. vapor, $s_g$ |
| 0.01 | 0.61169 | 0.001091 | 205.99 | −333.40 | 2707.9 | 2374.5 | −333.40 | 2833.9 | 2500.5 | −1.2202 | 10.374 | 9.154 |
| 0 | 0.61115 | 0.001091 | 206.17 | −333.43 | 2707.9 | 2374.5 | −333.43 | 2833.9 | 2500.5 | −1.2204 | 10.375 | 9.154 |
| −2 | 0.51772 | 0.001091 | 241.62 | −337.63 | 2709.4 | 2371.8 | −337.63 | 2834.5 | 2496.8 | −1.2358 | 10.453 | 9.218 |
| −4 | 0.43748 | 0.001090 | 283.84 | −341.80 | 2710.8 | 2369.0 | −341.80 | 2835.0 | 2493.2 | −1.2513 | 10.533 | 9.282 |
| −6 | 0.36873 | 0.001090 | 334.27 | −345.94 | 2712.2 | 2366.2 | −345.93 | 2835.4 | 2489.5 | −1.2667 | 10.613 | 9.347 |
| −8 | 0.30998 | 0.001090 | 394.66 | −350.04 | 2713.5 | 2363.5 | −350.04 | 2835.8 | 2485.8 | −1.2821 | 10.695 | 9.413 |
| −10 | 0.25990 | 0.001089 | 467.17 | −354.12 | 2714.8 | 2360.7 | −354.12 | 2836.2 | 2482.1 | −1.2976 | 10.778 | 9.480 |
| −12 | 0.21732 | 0.001089 | 554.47 | −358.17 | 2716.1 | 2357.9 | −358.17 | 2836.6 | 2478.4 | −1.3130 | 10.862 | 9.549 |
| −14 | 0.18121 | 0.001088 | 659.88 | −362.18 | 2717.3 | 2355.2 | −362.18 | 2836.9 | 2474.7 | −1.3284 | 10.947 | 9.618 |
| −16 | 0.15068 | 0.001088 | 787.51 | −366.17 | 2718.6 | 2352.4 | −366.17 | 2837.2 | 2471.0 | −1.3439 | 11.033 | 9.689 |
| −18 | 0.12492 | 0.001088 | 942.51 | −370.13 | 2719.7 | 2349.6 | −370.13 | 2837.5 | 2467.3 | −1.3593 | 11.121 | 9.761 |
| −20 | 0.10326 | 0.001087 | 1131.3 | −374.06 | 2720.9 | 2346.8 | −374.06 | 2837.7 | 2463.6 | −1.3748 | 11.209 | 9.835 |
| −22 | 0.08510 | 0.001087 | 1362.0 | −377.95 | 2722.0 | 2344.1 | −377.95 | 2837.9 | 2459.9 | −1.3903 | 11.300 | 9.909 |
| −24 | 0.06991 | 0.001087 | 1644.7 | −381.82 | 2723.1 | 2341.3 | −381.82 | 2838.1 | 2456.2 | −1.4057 | 11.391 | 9.985 |
| −26 | 0.05725 | 0.001087 | 1992.2 | −385.66 | 2724.2 | 2338.5 | −385.66 | 2838.2 | 2452.5 | −1.4212 | 11.484 | 10.063 |
| −28 | 0.04673 | 0.001086 | 2421.0 | −389.47 | 2725.2 | 2335.7 | −389.47 | 2838.3 | 2448.8 | −1.4367 | 11.578 | 10.141 |
| −30 | 0.03802 | 0.001086 | 2951.7 | −393.25 | 2726.2 | 2332.9 | −393.25 | 2838.4 | 2445.1 | −1.4521 | 11.673 | 10.221 |
| −32 | 0.03082 | 0.001086 | 3610.9 | −397.00 | 2727.2 | 2330.2 | −397.00 | 2838.4 | 2441.4 | −1.4676 | 11.770 | 10.303 |
| −34 | 0.02490 | 0.001085 | 4432.4 | −400.72 | 2728.1 | 2327.4 | −400.72 | 2838.5 | 2437.7 | −1.4831 | 11.869 | 10.386 |
| −36 | 0.02004 | 0.001085 | 5460.1 | −404.40 | 2729.0 | 2324.6 | −404.40 | 2838.4 | 2434.0 | −1.4986 | 11.969 | 10.470 |
| −38 | 0.01608 | 0.001085 | 6750.5 | −408.07 | 2729.9 | 2321.8 | −408.07 | 2838.4 | 2430.3 | −1.5141 | 12.071 | 10.557 |
| −40 | 0.01285 | 0.001084 | 8376.7 | −411.70 | 2730.7 | 2319.0 | −411.70 | 2838.3 | 2426.6 | −1.5296 | 12.174 | 10.644 |

**FIGURE A–9**

*T-s* diagram for water.

*Source of Data: From* NBS/NRC Steam Tables/1 *by Lester Haar, John S. Gallagher, and George S. Kell. Routledge/Taylor & Francis Books, Inc., 1984.*

**FIGURE A–10**
Mollier diagram for water.

*Source of Data: From* NBS/NRC Steam Tables/1 *by Lester Haar, John S. Gallagher, and George S. Kell. Routledge/Taylor & Francis Books, Inc.,* 1984.

## TABLE A–11

Saturated refrigerant-134a—Temperature table

| Temp., $T$ °C | Sat. press., $P_{sat}$ kPa | Specific volume, m³/kg | | Internal energy, kJ/kg | | | Enthalpy, kJ/kg | | | Entropy, kJ/kg·K | | |
|---|---|---|---|---|---|---|---|---|---|---|---|---|
| | | Sat. liquid, $v_f$ | Sat. vapor, $v_g$ | Sat. liquid, $u_f$ | Evap., $u_{fg}$ | Sat. vapor, $u_g$ | Sat. liquid, $h_f$ | Evap., $h_{fg}$ | Sat. vapor, $h_g$ | Sat. liquid, $s_f$ | Evap., $s_{fg}$ | Sat. vapor, $s_g$ |
| −40 | 51.25 | 0.0007053 | 0.36064 | −0.036 | 207.42 | 207.38 | 0.00 | 225.86 | 225.86 | 0.00000 | 0.96869 | 0.96869 |
| −38 | 56.86 | 0.0007082 | 0.32718 | 2.472 | 206.06 | 208.53 | 2.512 | 224.62 | 227.13 | 0.01071 | 0.95516 | 0.96588 |
| −36 | 62.95 | 0.0007111 | 0.29740 | 4.987 | 204.69 | 209.68 | 5.032 | 223.37 | 228.40 | 0.02137 | 0.94182 | 0.96319 |
| −34 | 69.56 | 0.0007141 | 0.27082 | 7.509 | 203.32 | 210.83 | 7.559 | 222.10 | 229.66 | 0.03196 | 0.92867 | 0.96063 |
| −32 | 76.71 | 0.0007171 | 0.24706 | 10.04 | 201.94 | 211.97 | 10.09 | 220.83 | 230.93 | 0.04249 | 0.91569 | 0.95819 |
| −30 | 84.43 | 0.0007201 | 0.22577 | 12.58 | 200.55 | 213.12 | 12.64 | 219.55 | 232.19 | 0.05297 | 0.90289 | 0.95586 |
| −28 | 92.76 | 0.0007232 | 0.20666 | 15.12 | 199.15 | 214.27 | 15.19 | 218.25 | 233.44 | 0.06339 | 0.89024 | 0.95364 |
| −26 | 101.73 | 0.0007264 | 0.18947 | 17.67 | 197.75 | 215.42 | 17.75 | 216.95 | 234.70 | 0.07376 | 0.87776 | 0.95152 |
| −24 | 111.37 | 0.0007296 | 0.17398 | 20.23 | 196.34 | 216.57 | 20.31 | 215.63 | 235.94 | 0.08408 | 0.86542 | 0.94950 |
| −22 | 121.72 | 0.0007328 | 0.15999 | 22.80 | 194.92 | 217.71 | 22.89 | 214.30 | 237.19 | 0.09435 | 0.85323 | 0.94758 |
| −20 | 132.82 | 0.0007361 | 0.14735 | 25.37 | 193.49 | 218.86 | 25.47 | 212.96 | 238.43 | 0.10456 | 0.84119 | 0.94575 |
| −18 | 144.69 | 0.0007394 | 0.13589 | 27.96 | 192.05 | 220.00 | 28.07 | 211.60 | 239.67 | 0.11473 | 0.82927 | 0.94401 |
| −16 | 157.38 | 0.0007428 | 0.12550 | 30.55 | 190.60 | 221.15 | 30.67 | 210.23 | 240.90 | 0.12486 | 0.81749 | 0.94234 |
| −14 | 170.93 | 0.0007463 | 0.11605 | 33.15 | 189.14 | 222.29 | 33.28 | 208.84 | 242.12 | 0.13493 | 0.80583 | 0.94076 |
| −12 | 185.37 | 0.0007498 | 0.10744 | 35.76 | 187.66 | 223.42 | 35.90 | 207.44 | 243.34 | 0.14497 | 0.79429 | 0.93925 |
| −10 | 200.74 | 0.0007533 | 0.099600 | 38.38 | 186.18 | 224.56 | 38.53 | 206.02 | 244.55 | 0.15496 | 0.78286 | 0.93782 |
| −8 | 217.08 | 0.0007570 | 0.092438 | 41.01 | 184.69 | 225.69 | 41.17 | 204.59 | 245.76 | 0.16491 | 0.77154 | 0.93645 |
| −6 | 234.44 | 0.0007607 | 0.085888 | 43.64 | 183.18 | 226.82 | 43.82 | 203.14 | 246.95 | 0.17482 | 0.76033 | 0.93514 |
| −4 | 252.85 | 0.0007644 | 0.079889 | 46.29 | 181.66 | 227.94 | 46.48 | 201.66 | 248.14 | 0.18469 | 0.74921 | 0.93390 |
| −2 | 272.36 | 0.0007683 | 0.074388 | 48.94 | 180.12 | 229.07 | 49.15 | 200.17 | 249.33 | 0.19452 | 0.73819 | 0.93271 |
| 0 | 293.01 | 0.0007722 | 0.069335 | 51.61 | 178.58 | 230.18 | 51.83 | 198.67 | 250.50 | 0.20432 | 0.72726 | 0.93158 |
| 2 | 314.84 | 0.0007761 | 0.064690 | 54.28 | 177.01 | 231.30 | 54.53 | 197.14 | 251.66 | 0.21408 | 0.71641 | 0.93050 |
| 4 | 337.90 | 0.0007802 | 0.060412 | 56.97 | 175.44 | 232.40 | 57.23 | 195.58 | 252.82 | 0.22381 | 0.70565 | 0.92946 |
| 6 | 362.23 | 0.0007843 | 0.056469 | 59.66 | 173.84 | 233.51 | 59.95 | 194.01 | 253.96 | 0.23351 | 0.69496 | 0.92847 |
| 8 | 387.88 | 0.0007886 | 0.052829 | 62.37 | 172.23 | 234.60 | 62.68 | 192.42 | 255.09 | 0.24318 | 0.68435 | 0.92752 |
| 10 | 414.89 | 0.0007929 | 0.049466 | 65.09 | 170.61 | 235.69 | 65.42 | 190.80 | 256.22 | 0.25282 | 0.67380 | 0.92661 |
| 12 | 443.31 | 0.0007973 | 0.046354 | 67.82 | 168.96 | 236.78 | 68.17 | 189.16 | 257.33 | 0.26243 | 0.66331 | 0.92574 |
| 14 | 473.19 | 0.0008018 | 0.043471 | 70.56 | 167.30 | 237.86 | 70.94 | 187.49 | 258.43 | 0.27201 | 0.65289 | 0.92490 |
| 16 | 504.58 | 0.0008064 | 0.040798 | 73.31 | 165.62 | 238.93 | 73.72 | 185.80 | 259.51 | 0.28157 | 0.64252 | 0.92409 |
| 18 | 537.52 | 0.0008112 | 0.038317 | 76.07 | 163.92 | 239.99 | 76.51 | 184.08 | 260.59 | 0.29111 | 0.63219 | 0.92330 |
| 20 | 572.07 | 0.0008160 | 0.036012 | 78.85 | 162.19 | 241.04 | 79.32 | 182.33 | 261.64 | 0.30062 | 0.62192 | 0.92254 |
| 22 | 608.27 | 0.0008209 | 0.033867 | 81.64 | 160.45 | 242.09 | 82.14 | 180.55 | 262.69 | 0.31012 | 0.61168 | 0.92180 |
| 24 | 646.18 | 0.0008260 | 0.031869 | 84.44 | 158.68 | 243.13 | 84.98 | 178.74 | 263.72 | 0.31959 | 0.60148 | 0.92107 |
| 26 | 685.84 | 0.0008312 | 0.030008 | 87.26 | 156.89 | 244.15 | 87.83 | 176.90 | 264.73 | 0.32905 | 0.59131 | 0.92036 |
| 28 | 727.31 | 0.0008366 | 0.028271 | 90.09 | 155.08 | 245.17 | 90.70 | 175.03 | 265.73 | 0.33849 | 0.58117 | 0.91967 |
| 30 | 770.64 | 0.0008421 | 0.026648 | 92.93 | 153.24 | 246.17 | 93.58 | 173.13 | 266.71 | 0.34792 | 0.57105 | 0.91897 |
| 32 | 815.89 | 0.0008477 | 0.025131 | 95.79 | 151.37 | 247.17 | 96.49 | 171.19 | 267.67 | 0.35734 | 0.56095 | 0.91829 |
| 34 | 863.11 | 0.0008535 | 0.023712 | 98.67 | 149.48 | 248.15 | 99.41 | 169.21 | 268.61 | 0.36675 | 0.55086 | 0.91760 |
| 36 | 912.35 | 0.0008595 | 0.022383 | 101.56 | 147.55 | 249.11 | 102.34 | 167.19 | 269.53 | 0.37615 | 0.54077 | 0.91692 |
| 38 | 963.68 | 0.0008657 | 0.021137 | 104.47 | 145.60 | 250.07 | 105.30 | 165.13 | 270.44 | 0.38554 | 0.53068 | 0.91622 |
| 40 | 1017.1 | 0.0008720 | 0.019968 | 107.39 | 143.61 | 251.00 | 108.28 | 163.03 | 271.31 | 0.39493 | 0.52059 | 0.91552 |
| 42 | 1072.8 | 0.0008786 | 0.018870 | 110.34 | 141.59 | 251.92 | 111.28 | 160.89 | 272.17 | 0.40432 | 0.51048 | 0.91480 |
| 44 | 1130.7 | 0.0008854 | 0.017837 | 113.30 | 139.53 | 252.83 | 114.30 | 158.70 | 273.00 | 0.41371 | 0.50036 | 0.91407 |

Saturated refrigerant-134a—Temperature table *(Concluded)*

| Temp., $T$ °C | Sat. press., $P_{sat}$ kPa | Specific volume, m³/kg | | Internal energy, kJ/kg | | | Enthalpy, kJ/kg | | | Entropy, kJ/kg·K | | |
|---|---|---|---|---|---|---|---|---|---|---|---|---|
| | | Sat. liquid, $v_f$ | Sat. vapor, $v_g$ | Sat. liquid, $u_f$ | Evap., $u_{fg}$ | Sat. vapor, $u_g$ | Sat. liquid, $h_f$ | Evap., $h_{fg}$ | Sat. vapor, $h_g$ | Sat. liquid, $s_f$ | Evap., $s_{fg}$ | Sat. vapor, $s_g$ |
| 46 | 1191.0 | 0.0008924 | 0.016866 | 116.28 | 137.43 | 253.71 | 117.34 | 156.46 | 273.80 | 0.42311 | 0.49020 | 0.91331 |
| 48 | 1253.6 | 0.0008997 | 0.015951 | 119.28 | 135.30 | 254.58 | 120.41 | 154.17 | 274.57 | 0.43251 | 0.48001 | 0.91252 |
| 52 | 1386.2 | 0.0009151 | 0.014276 | 125.35 | 130.89 | 256.24 | 126.62 | 149.41 | 276.03 | 0.45136 | 0.45948 | 0.91084 |
| 56 | 1529.1 | 0.0009317 | 0.012782 | 131.52 | 126.29 | 257.81 | 132.94 | 144.41 | 277.35 | 0.47028 | 0.43870 | 0.90898 |
| 60 | 1682.8 | 0.0009498 | 0.011434 | 137.79 | 121.45 | 259.23 | 139.38 | 139.09 | 278.47 | 0.48930 | 0.41746 | 0.90676 |
| 65 | 1891.0 | 0.0009751 | 0.009959 | 145.80 | 115.06 | 260.86 | 147.64 | 132.05 | 279.69 | 0.51330 | 0.39048 | 0.90379 |
| 70 | 2118.2 | 0.0010037 | 0.008650 | 154.03 | 108.17 | 262.20 | 156.15 | 124.37 | 280.52 | 0.53763 | 0.36239 | 0.90002 |
| 75 | 2365.8 | 0.0010373 | 0.007486 | 162.55 | 100.62 | 263.17 | 165.01 | 115.87 | 280.88 | 0.56252 | 0.33279 | 0.89531 |
| 80 | 2635.3 | 0.0010774 | 0.006439 | 171.43 | 92.22 | 263.66 | 174.27 | 106.35 | 280.63 | 0.58812 | 0.30113 | 0.88925 |
| 85 | 2928.2 | 0.0011273 | 0.005484 | 180.81 | 82.64 | 263.45 | 184.11 | 95.39 | 279.51 | 0.61487 | 0.26632 | 0.88120 |
| 90 | 3246.9 | 0.0011938 | 0.004591 | 190.94 | 71.19 | 262.13 | 194.82 | 82.22 | 277.04 | 0.64354 | 0.22638 | 0.86991 |
| 95 | 3594.1 | 0.0012945 | 0.003713 | 202.49 | 56.25 | 258.73 | 207.14 | 64.94 | 272.08 | 0.67605 | 0.17638 | 0.85243 |
| 100 | 3975.1 | 0.0015269 | 0.002657 | 218.73 | 29.72 | 248.46 | 224.80 | 34.22 | 259.02 | 0.72224 | 0.09169 | 0.81393 |

*Source of Data:* Tables A–11 through A–13 are generated using the Engineering Equation Solver (EES) software developed by S. A. Klein and F. L. Alvarado. The routine used in calculations is the R134a, which is based on the fundamental equation of state developed by R. Tillner–Roth and H.D. Baehr, "An International Standard Formulation for the Thermodynamic Properties of 1,1,1,2-Tetrafluoroethane (HFC-134a) for temperatures from 170 K to 455 K and pressures up to 70 MPa," *J. Phys. Chem, Ref. Data,* Vol. 23, No. 5, 1994. The enthalpy and entropy values of saturated liquid are set to zero at −40°C (and −40°F).

## TABLE A–12

Saturated refrigerant-134a—Pressure table

| Press., $P$ kPa | Sat. temp., $T_{sat}$ °C | Specific volume, m³/kg | | Internal energy, kJ/kg | | | Enthalpy, kJ/kg | | | Entropy, kJ/kg·K | | |
|---|---|---|---|---|---|---|---|---|---|---|---|---|
| | | Sat. liquid, $v_f$ | Sat. vapor, $v_g$ | Sat. liquid, $u_f$ | Evap., $u_{fg}$ | Sat. vapor, $u_g$ | Sat. liquid, $h_f$ | Evap., $h_{fg}$ | Sat. vapor, $h_g$ | Sat. liquid, $s_f$ | Evap., $s_{fg}$ | Sat. vapor, $s_g$ |
| 60 | −36.95 | 0.0007097 | 0.31108 | 3.795 | 205.34 | 209.13 | 3.837 | 223.96 | 227.80 | 0.01633 | 0.94812 | 0.96445 |
| 70 | −33.87 | 0.0007143 | 0.26921 | 7.672 | 203.23 | 210.90 | 7.722 | 222.02 | 229.74 | 0.03264 | 0.92783 | 0.96047 |
| 80 | −31.13 | 0.0007184 | 0.23749 | 11.14 | 201.33 | 212.48 | 11.20 | 220.27 | 231.47 | 0.04707 | 0.91009 | 0.95716 |
| 90 | −28.65 | 0.0007222 | 0.21261 | 14.30 | 199.60 | 213.90 | 14.36 | 218.67 | 233.04 | 0.06003 | 0.89431 | 0.95434 |
| 100 | −26.37 | 0.0007258 | 0.19255 | 17.19 | 198.01 | 215.21 | 17.27 | 217.19 | 234.46 | 0.07182 | 0.88008 | 0.95191 |
| 120 | −22.32 | 0.0007323 | 0.16216 | 22.38 | 195.15 | 217.53 | 22.47 | 214.52 | 236.99 | 0.09269 | 0.85520 | 0.94789 |
| 140 | −18.77 | 0.0007381 | 0.14020 | 26.96 | 192.60 | 219.56 | 27.06 | 212.13 | 239.19 | 0.11080 | 0.83387 | 0.94467 |
| 160 | −15.60 | 0.0007435 | 0.12355 | 31.06 | 190.31 | 221.37 | 31.18 | 209.96 | 241.14 | 0.12686 | 0.81517 | 0.94202 |
| 180 | −12.73 | 0.0007485 | 0.11049 | 34.81 | 188.20 | 223.01 | 34.94 | 207.95 | 242.90 | 0.14131 | 0.79848 | 0.93979 |
| 200 | −10.09 | 0.0007532 | 0.099951 | 38.26 | 186.25 | 224.51 | 38.41 | 206.09 | 244.50 | 0.15449 | 0.78339 | 0.93788 |
| 240 | −5.38 | 0.0007618 | 0.083983 | 44.46 | 182.71 | 227.17 | 44.64 | 202.68 | 247.32 | 0.17786 | 0.75689 | 0.93475 |
| 280 | −1.25 | 0.0007697 | 0.072434 | 49.95 | 179.54 | 229.49 | 50.16 | 199.61 | 249.77 | 0.19822 | 0.73406 | 0.93228 |
| 320 | 2.46 | 0.0007771 | 0.063681 | 54.90 | 176.65 | 231.55 | 55.14 | 196.78 | 251.93 | 0.21631 | 0.71395 | 0.93026 |
| 360 | 5.82 | 0.0007840 | 0.056809 | 59.42 | 173.99 | 233.41 | 59.70 | 194.15 | 253.86 | 0.23265 | 0.69591 | 0.92856 |
| 400 | 8.91 | 0.0007905 | 0.051266 | 63.61 | 171.49 | 235.10 | 63.92 | 191.68 | 255.61 | 0.24757 | 0.67954 | 0.92711 |
| 450 | 12.46 | 0.0007983 | 0.045677 | 68.44 | 168.58 | 237.03 | 68.80 | 188.78 | 257.58 | 0.26462 | 0.66093 | 0.92555 |
| 500 | 15.71 | 0.0008058 | 0.041168 | 72.92 | 165.86 | 238.77 | 73.32 | 186.04 | 259.36 | 0.28021 | 0.64399 | 0.92420 |
| 550 | 18.73 | 0.0008129 | 0.037452 | 77.09 | 163.29 | 240.38 | 77.54 | 183.44 | 260.98 | 0.29460 | 0.62842 | 0.92302 |
| 600 | 21.55 | 0.0008198 | 0.034335 | 81.01 | 160.84 | 241.86 | 81.50 | 180.95 | 262.46 | 0.30799 | 0.61398 | 0.92196 |
| 650 | 24.20 | 0.0008265 | 0.031680 | 84.72 | 158.51 | 243.23 | 85.26 | 178.56 | 263.82 | 0.32052 | 0.60048 | 0.92100 |
| 700 | 26.69 | 0.0008331 | 0.029392 | 88.24 | 156.27 | 244.51 | 88.82 | 176.26 | 265.08 | 0.33232 | 0.58780 | 0.92012 |
| 750 | 29.06 | 0.0008395 | 0.027398 | 91.59 | 154.11 | 245.70 | 92.22 | 174.03 | 266.25 | 0.34348 | 0.57582 | 0.91930 |
| 800 | 31.31 | 0.0008457 | 0.025645 | 94.80 | 152.02 | 246.82 | 95.48 | 171.86 | 267.34 | 0.35408 | 0.56445 | 0.91853 |
| 850 | 33.45 | 0.0008519 | 0.024091 | 97.88 | 150.00 | 247.88 | 98.61 | 169.75 | 268.36 | 0.36417 | 0.55362 | 0.91779 |
| 900 | 35.51 | 0.0008580 | 0.022703 | 100.84 | 148.03 | 248.88 | 101.62 | 167.69 | 269.31 | 0.37383 | 0.54326 | 0.91709 |
| 950 | 37.48 | 0.0008640 | 0.021456 | 103.70 | 146.11 | 249.82 | 104.52 | 165.68 | 270.20 | 0.38307 | 0.53333 | 0.91641 |
| 1000 | 39.37 | 0.0008700 | 0.020329 | 106.47 | 144.24 | 250.71 | 107.34 | 163.70 | 271.04 | 0.39196 | 0.52378 | 0.91574 |
| 1200 | 46.29 | 0.0008935 | 0.016728 | 116.72 | 137.12 | 253.84 | 117.79 | 156.12 | 273.92 | 0.42449 | 0.48870 | 0.91320 |
| 1400 | 52.40 | 0.0009167 | 0.014119 | 125.96 | 130.44 | 256.40 | 127.25 | 148.92 | 276.17 | 0.45325 | 0.45742 | 0.91067 |
| 1600 | 57.88 | 0.0009400 | 0.012134 | 134.45 | 124.05 | 258.50 | 135.96 | 141.96 | 277.92 | 0.47921 | 0.42881 | 0.90802 |
| 1800 | 62.87 | 0.0009639 | 0.010568 | 142.36 | 117.85 | 260.21 | 144.09 | 135.14 | 279.23 | 0.50304 | 0.40213 | 0.90517 |
| 2000 | 67.45 | 0.0009887 | 0.009297 | 149.81 | 111.75 | 261.56 | 151.78 | 128.36 | 280.15 | 0.52519 | 0.37684 | 0.90204 |
| 2500 | 77.54 | 0.0010567 | 0.006941 | 167.02 | 96.47 | 263.49 | 169.66 | 111.18 | 280.84 | 0.57542 | 0.31701 | 0.89243 |
| 3000 | 86.16 | 0.0011410 | 0.005272 | 183.09 | 80.17 | 263.26 | 186.51 | 92.57 | 279.08 | 0.62133 | 0.25759 | 0.87893 |

## TABLE A–13

Superheated refrigerant-134a

| $T$ °C | $v$ m³/kg | $u$ kJ/kg | $h$ kJ/kg | $s$ kJ/kg·K | $v$ m³/kg | $u$ kJ/kg | $h$ kJ/kg | $s$ kJ/kg·K | $v$ m³/kg | $u$ kJ/kg | $h$ kJ/kg | $s$ kJ/kg·K |
|---|---|---|---|---|---|---|---|---|---|---|---|---|
| | $P = 0.06$ MPa ($T_{sat} = -36.95$°C) | | | | $P = 0.10$ MPa ($T_{sat} = -26.37$°C) | | | | $P = 0.14$ MPa ($T_{sat} = -18.77$°C) | | | |
| Sat. | 0.31108 | 209.13 | 227.80 | 0.9645 | 0.19255 | 215.21 | 234.46 | 0.9519 | 0.14020 | 219.56 | 239.19 | 0.9447 |
| −20 | 0.33608 | 220.62 | 240.78 | 1.0175 | 0.19841 | 219.68 | 239.52 | 0.9721 | | | | |
| −10 | 0.35048 | 227.57 | 248.60 | 1.0478 | 0.20743 | 226.77 | 247.51 | 1.0031 | 0.14605 | 225.93 | 246.37 | 0.9724 |
| 0 | 0.36476 | 234.67 | 256.56 | 1.0775 | 0.21630 | 233.97 | 255.60 | 1.0333 | 0.15263 | 233.25 | 254.61 | 1.0032 |
| 10 | 0.37893 | 241.94 | 264.68 | 1.1067 | 0.22506 | 241.32 | 263.82 | 1.0628 | 0.15908 | 240.68 | 262.95 | 1.0331 |
| 20 | 0.39302 | 249.37 | 272.95 | 1.1354 | 0.23373 | 248.81 | 272.18 | 1.0919 | 0.16544 | 248.24 | 271.40 | 1.0625 |
| 30 | 0.40705 | 256.97 | 281.39 | 1.1637 | 0.24233 | 256.46 | 280.69 | 1.1204 | 0.17172 | 255.95 | 279.99 | 1.0913 |
| 40 | 0.42102 | 264.73 | 289.99 | 1.1916 | 0.25088 | 264.27 | 289.36 | 1.1485 | 0.17794 | 263.80 | 288.72 | 1.1196 |
| 50 | 0.43495 | 272.66 | 298.75 | 1.2192 | 0.25937 | 272.24 | 298.17 | 1.1762 | 0.18412 | 271.81 | 297.59 | 1.1475 |
| 60 | 0.44883 | 280.75 | 307.68 | 1.2464 | 0.26783 | 280.36 | 307.15 | 1.2036 | 0.19025 | 279.97 | 306.61 | 1.1750 |
| 70 | 0.46269 | 289.01 | 316.77 | 1.2732 | 0.27626 | 288.65 | 316.28 | 1.2306 | 0.19635 | 288.29 | 315.78 | 1.2021 |
| 80 | 0.47651 | 297.43 | 326.02 | 1.2998 | 0.28465 | 297.10 | 325.57 | 1.2573 | 0.20242 | 296.77 | 325.11 | 1.2289 |
| 90 | 0.49032 | 306.02 | 335.43 | 1.3261 | 0.29303 | 305.71 | 335.01 | 1.2836 | 0.20847 | 305.40 | 334.59 | 1.2554 |
| 100 | 0.50410 | 314.76 | 345.01 | 1.3521 | 0.30138 | 314.48 | 344.61 | 1.3097 | 0.21449 | 314.19 | 344.22 | 1.2815 |
| | $P = 0.18$ MPa ($T_{sat} = -12.73$°C) | | | | $P = 0.20$ MPa ($T_{sat} = -10.09$°C) | | | | $P = 0.24$ MPa ($T_{sat} = -5.38$°C) | | | |
| Sat. | 0.11049 | 223.01 | 242.90 | 0.9398 | 0.09995 | 224.51 | 244.50 | 0.9379 | 0.08398 | 227.17 | 247.32 | 0.9348 |
| −10 | 0.11189 | 225.04 | 245.18 | 0.9485 | 0.09991 | 224.57 | 244.56 | 0.9381 | | | | |
| 0 | 0.11722 | 232.49 | 253.59 | 0.9799 | 0.10481 | 232.11 | 253.07 | 0.9699 | 0.08617 | 231.30 | 251.98 | 0.9520 |
| 10 | 0.12240 | 240.02 | 262.05 | 1.0103 | 0.10955 | 239.69 | 261.60 | 1.0005 | 0.09026 | 239.00 | 260.66 | 0.9832 |
| 20 | 0.12748 | 247.66 | 270.60 | 1.0400 | 0.11418 | 247.36 | 270.20 | 1.0304 | 0.09423 | 246.76 | 269.38 | 1.0134 |
| 30 | 0.13248 | 255.43 | 279.27 | 1.0691 | 0.11874 | 255.16 | 278.91 | 1.0596 | 0.09812 | 254.63 | 278.17 | 1.0429 |
| 40 | 0.13741 | 263.33 | 288.07 | 1.0976 | 0.12322 | 263.09 | 287.74 | 1.0882 | 0.10193 | 262.61 | 287.07 | 1.0718 |
| 50 | 0.14230 | 271.38 | 297.00 | 1.1257 | 0.12766 | 271.16 | 296.70 | 1.1164 | 0.10570 | 270.73 | 296.09 | 1.1002 |
| 60 | 0.14715 | 279.58 | 306.07 | 1.1533 | 0.13206 | 279.38 | 305.79 | 1.1441 | 0.10942 | 278.98 | 305.24 | 1.1281 |
| 70 | 0.15196 | 287.93 | 315.28 | 1.1806 | 0.13641 | 287.75 | 315.03 | 1.1714 | 0.11310 | 287.38 | 314.53 | 1.1555 |
| 80 | 0.15673 | 296.43 | 324.65 | 1.2075 | 0.14074 | 296.27 | 324.41 | 1.1984 | 0.11675 | 295.93 | 323.95 | 1.1826 |
| 90 | 0.16149 | 305.09 | 334.16 | 1.2340 | 0.14504 | 304.93 | 333.94 | 1.2250 | 0.12038 | 304.62 | 333.51 | 1.2093 |
| 100 | 0.16622 | 313.90 | 343.82 | 1.2603 | 0.14933 | 313.75 | 343.62 | 1.2513 | 0.12398 | 313.46 | 343.22 | 1.2356 |
| | $P = 0.28$ MPa ($T_{sat} = -1.25$°C) | | | | $P = 0.32$ MPa ($T_{sat} = 2.46$°C) | | | | $P = 0.40$ MPa ($T_{sat} = 8.91$°C) | | | |
| Sat. | 0.07243 | 229.49 | 249.77 | 0.9323 | 0.06368 | 231.55 | 251.93 | 0.9303 | 0.051266 | 235.10 | 255.61 | 0.9271 |
| 0 | 0.07282 | 230.46 | 250.85 | 0.9362 | | | | | | | | |
| 10 | 0.07646 | 238.29 | 259.70 | 0.9681 | 0.06609 | 237.56 | 258.70 | 0.9545 | 0.051506 | 235.99 | 256.59 | 0.9306 |
| 20 | 0.07997 | 246.15 | 268.54 | 0.9987 | 0.06925 | 245.51 | 267.67 | 0.9856 | 0.054213 | 244.19 | 265.88 | 0.9628 |
| 30 | 0.08338 | 254.08 | 277.42 | 1.0285 | 0.07231 | 253.52 | 276.66 | 1.0158 | 0.056796 | 252.37 | 275.09 | 0.9937 |
| 40 | 0.08672 | 262.12 | 286.40 | 1.0577 | 0.07530 | 261.62 | 285.72 | 1.0452 | 0.059292 | 260.60 | 284.32 | 1.0237 |
| 50 | 0.09000 | 270.28 | 295.48 | 1.0862 | 0.07823 | 269.83 | 294.87 | 1.0739 | 0.061724 | 268.92 | 293.61 | 1.0529 |
| 60 | 0.09324 | 278.58 | 304.69 | 1.1143 | 0.08111 | 278.17 | 304.12 | 1.1022 | 0.064104 | 277.34 | 302.98 | 1.0814 |
| 70 | 0.09644 | 287.01 | 314.01 | 1.1419 | 0.08395 | 286.64 | 313.50 | 1.1299 | 0.066443 | 285.88 | 312.45 | 1.1095 |
| 80 | 0.09961 | 295.59 | 323.48 | 1.1690 | 0.08675 | 295.24 | 323.00 | 1.1572 | 0.068747 | 294.54 | 322.04 | 1.1370 |
| 90 | 0.10275 | 304.30 | 333.07 | 1.1958 | 0.08953 | 303.99 | 332.64 | 1.1841 | 0.071023 | 303.34 | 331.75 | 1.1641 |
| 100 | 0.10587 | 313.17 | 342.81 | 1.2223 | 0.09229 | 312.87 | 342.41 | 1.2106 | 0.073274 | 312.28 | 341.59 | 1.1908 |
| 110 | 0.10897 | 322.18 | 352.69 | 1.2484 | 0.09503 | 321.91 | 352.31 | 1.2368 | 0.075504 | 321.35 | 351.55 | 1.2172 |
| 120 | 0.11205 | 331.34 | 362.72 | 1.2742 | 0.09775 | 331.08 | 362.36 | 1.2627 | 0.077717 | 330.56 | 361.65 | 1.2432 |
| 130 | 0.11512 | 340.65 | 372.88 | 1.2998 | 0.10045 | 340.41 | 372.55 | 1.2883 | 0.079913 | 339.92 | 371.89 | 1.2689 |
| 140 | 0.11818 | 350.11 | 383.20 | 1.3251 | 0.10314 | 349.88 | 382.89 | 1.3136 | 0.082096 | 349.42 | 382.26 | 1.2943 |

Superheated refrigerant-134a *(Concluded)*

| T °C | υ m³/kg | u kJ/kg | h kJ/kg | s kJ/kg·K | υ m³/kg | u kJ/kg | h kJ/kg | s kJ/kg·K | υ m³/kg | u kJ/kg | h kJ/kg | s kJ/kg·K |
|---|---|---|---|---|---|---|---|---|---|---|---|---|
| | $P = 0.50$ MPa ($T_{sat} = 15.71$°C) | | | | $P = 0.60$ MPa ($T_{sat} = 21.55$°C) | | | | $P = 0.70$ MPa ($T_{sat} = 26.69$°C) | | | |
| Sat. | 0.041168 | 238.77 | 259.36 | 0.9242 | 0.034335 | 241.86 | 262.46 | 0.9220 | 0.029392 | 244.51 | 265.08 | 0.9201 |
| 20 | 0.042115 | 242.42 | 263.48 | 0.9384 | | | | | | | | |
| 30 | 0.044338 | 250.86 | 273.03 | 0.9704 | 0.035984 | 249.24 | 270.83 | 0.9500 | 0.029966 | 247.49 | 268.47 | 0.9314 |
| 40 | 0.046456 | 259.27 | 282.50 | 1.0011 | 0.037865 | 257.88 | 280.60 | 0.9817 | 0.031696 | 256.41 | 278.59 | 0.9642 |
| 50 | 0.048499 | 267.73 | 291.98 | 1.0309 | 0.039659 | 266.50 | 290.30 | 1.0122 | 0.033322 | 265.22 | 288.54 | 0.9955 |
| 60 | 0.050485 | 276.27 | 301.51 | 1.0600 | 0.041389 | 275.17 | 300.00 | 1.0417 | 0.034875 | 274.03 | 298.44 | 1.0257 |
| 70 | 0.052427 | 284.91 | 311.12 | 1.0884 | 0.043069 | 283.91 | 309.75 | 1.0706 | 0.036373 | 282.88 | 308.34 | 1.0550 |
| 80 | 0.054331 | 293.65 | 320.82 | 1.1163 | 0.044710 | 292.74 | 319.57 | 1.0988 | 0.037829 | 291.81 | 318.29 | 1.0835 |
| 90 | 0.056205 | 302.52 | 330.63 | 1.1436 | 0.046318 | 301.69 | 329.48 | 1.1265 | 0.039250 | 300.84 | 328.31 | 1.1115 |
| 100 | 0.058053 | 311.52 | 340.55 | 1.1706 | 0.047900 | 310.75 | 339.49 | 1.1536 | 0.040642 | 309.96 | 338.41 | 1.1389 |
| 110 | 0.059880 | 320.65 | 350.59 | 1.1971 | 0.049458 | 319.93 | 349.61 | 1.1804 | 0.042010 | 319.21 | 348.61 | 1.1659 |
| 120 | 0.061687 | 329.91 | 360.75 | 1.2233 | 0.050997 | 329.24 | 359.84 | 1.2068 | 0.043358 | 328.57 | 358.92 | 1.1925 |
| 130 | 0.063479 | 339.31 | 371.05 | 1.2492 | 0.052519 | 338.69 | 370.20 | 1.2328 | 0.044688 | 338.06 | 369.34 | 1.2186 |
| 140 | 0.065256 | 348.85 | 381.47 | 1.2747 | 0.054027 | 348.26 | 380.68 | 1.2585 | 0.046004 | 347.67 | 379.88 | 1.2445 |
| 150 | 0.067021 | 358.52 | 392.04 | 1.3000 | 0.055522 | 357.98 | 391.29 | 1.2838 | 0.047306 | 357.42 | 390.54 | 1.2700 |
| 160 | 0.068775 | 368.34 | 402.73 | 1.3250 | 0.057006 | 367.83 | 402.03 | 1.3089 | 0.048597 | 367.31 | 401.32 | 1.2952 |
| | $P = 0.80$ MPa ($T_{sat} = 31.31$°C) | | | | $P = 0.90$ MPa ($T_{sat} = 35.51$°C) | | | | $P = 1.00$ MPa ($T_{sat} = 39.37$°C) | | | |
| Sat. | 0.025645 | 246.82 | 267.34 | 0.9185 | 0.022686 | 248.82 | 269.25 | 0.9169 | 0.020319 | 250.71 | 271.04 | 0.9157 |
| 40 | 0.027035 | 254.84 | 276.46 | 0.9481 | 0.023375 | 253.15 | 274.19 | 0.9328 | 0.020406 | 251.32 | 271.73 | 0.9180 |
| 50 | 0.028547 | 263.87 | 286.71 | 0.9803 | 0.024809 | 262.46 | 284.79 | 0.9661 | 0.021796 | 260.96 | 282.76 | 0.9526 |
| 60 | 0.029973 | 272.85 | 296.82 | 1.0111 | 0.026146 | 271.62 | 295.15 | 0.9977 | 0.023068 | 270.33 | 293.40 | 0.9851 |
| 70 | 0.031340 | 281.83 | 306.90 | 1.0409 | 0.027413 | 280.74 | 305.41 | 1.0280 | 0.024261 | 279.61 | 303.87 | 1.0160 |
| 80 | 0.032659 | 290.86 | 316.99 | 1.0699 | 0.028630 | 289.88 | 315.65 | 1.0574 | 0.025398 | 288.87 | 314.27 | 1.0459 |
| 90 | 0.033941 | 299.97 | 327.12 | 1.0982 | 0.029806 | 299.08 | 325.90 | 1.0861 | 0.026492 | 298.17 | 324.66 | 1.0749 |
| 100 | 0.035193 | 309.17 | 337.32 | 1.1259 | 0.030951 | 308.35 | 336.21 | 1.1141 | 0.027552 | 307.52 | 335.08 | 1.1032 |
| 110 | 0.036420 | 318.47 | 347.61 | 1.1531 | 0.032068 | 317.72 | 346.58 | 1.1415 | 0.028584 | 316.96 | 345.54 | 1.1309 |
| 120 | 0.037625 | 327.89 | 357.99 | 1.1798 | 0.033164 | 327.19 | 357.04 | 1.1684 | 0.029592 | 326.49 | 356.08 | 1.1580 |
| 130 | 0.038813 | 337.42 | 368.47 | 1.2062 | 0.034241 | 336.78 | 367.59 | 1.1949 | 0.030581 | 336.12 | 366.70 | 1.1847 |
| 140 | 0.039985 | 347.08 | 379.07 | 1.2321 | 0.035302 | 346.48 | 378.25 | 1.2211 | 0.031554 | 345.87 | 377.42 | 1.2110 |
| 150 | 0.041143 | 356.86 | 389.78 | 1.2577 | 0.036349 | 356.30 | 389.01 | 1.2468 | 0.032512 | 355.73 | 388.24 | 1.2369 |
| 160 | 0.042290 | 366.78 | 400.61 | 1.2830 | 0.037384 | 366.25 | 399.89 | 1.2722 | 0.033457 | 365.71 | 399.17 | 1.2624 |
| 170 | 0.043427 | 376.83 | 411.57 | 1.3081 | 0.038408 | 376.33 | 410.89 | 1.2973 | 0.034392 | 375.82 | 410.22 | 1.2876 |
| 180 | 0.044554 | 387.01 | 422.65 | 1.3328 | 0.039423 | 386.54 | 422.02 | 1.3221 | 0.035317 | 386.06 | 421.38 | 1.3125 |
| | $P = 1.20$ MPa ($T_{sat} = 46.29$°C) | | | | $P = 1.40$ MPa ($T_{sat} = 52.40$°C) | | | | $P = 1.60$ MPa ($T_{sat} = 57.88$°C) | | | |
| Sat. | 0.016728 | 253.84 | 273.92 | 0.9132 | 0.014119 | 256.40 | 276.17 | 0.9107 | 0.012134 | 258.50 | 277.92 | 0.9080 |
| 50 | 0.017201 | 257.64 | 278.28 | 0.9268 | | | | | | | | |
| 60 | 0.018404 | 267.57 | 289.66 | 0.9615 | 0.015005 | 264.46 | 285.47 | 0.9389 | 0.012372 | 260.91 | 280.71 | 0.9164 |
| 70 | 0.019502 | 277.23 | 300.63 | 0.9939 | 0.016060 | 274.62 | 297.10 | 0.9733 | 0.013430 | 271.78 | 293.27 | 0.9536 |
| 80 | 0.020529 | 286.77 | 311.40 | 1.0249 | 0.017023 | 284.51 | 308.34 | 1.0056 | 0.014362 | 282.11 | 305.09 | 0.9875 |
| 90 | 0.021506 | 296.28 | 322.09 | 1.0547 | 0.017923 | 294.28 | 319.37 | 1.0364 | 0.015215 | 292.19 | 316.53 | 1.0195 |
| 100 | 0.022442 | 305.81 | 332.74 | 1.0836 | 0.018778 | 304.01 | 330.30 | 1.0661 | 0.016014 | 302.16 | 327.78 | 1.0501 |
| 110 | 0.023348 | 315.40 | 343.41 | 1.1119 | 0.019597 | 313.76 | 341.19 | 1.0949 | 0.016773 | 312.09 | 338.93 | 1.0795 |
| 120 | 0.024228 | 325.05 | 354.12 | 1.1395 | 0.020388 | 323.55 | 352.09 | 1.1230 | 0.017500 | 322.03 | 350.03 | 1.1081 |
| 130 | 0.025086 | 334.79 | 364.90 | 1.1665 | 0.021155 | 333.41 | 363.02 | 1.1504 | 0.018201 | 332.02 | 361.14 | 1.1360 |
| 140 | 0.025927 | 344.63 | 375.74 | 1.1931 | 0.021904 | 343.34 | 374.01 | 1.1773 | 0.018882 | 342.06 | 372.27 | 1.1633 |
| 150 | 0.026753 | 354.57 | 386.68 | 1.2192 | 0.022636 | 353.37 | 385.07 | 1.2038 | 0.019545 | 352.19 | 383.46 | 1.1901 |
| 160 | 0.027566 | 364.63 | 397.71 | 1.2450 | 0.023355 | 363.51 | 396.20 | 1.2298 | 0.020194 | 362.40 | 394.71 | 1.2164 |
| 170 | 0.028367 | 374.80 | 408.84 | 1.2704 | 0.024061 | 373.75 | 407.43 | 1.2554 | 0.020830 | 372.71 | 406.04 | 1.2422 |
| 180 | 0.029158 | 385.10 | 420.09 | 1.2955 | 0.024757 | 384.12 | 418.78 | 1.2808 | 0.021456 | 383.13 | 417.46 | 1.2677 |

**FIGURE A–14**

*P-h* diagram for refrigerant-134a.

*Note:* The reference point used for the chart is different than that used in the R-134a tables. Therefore, problems should be solved using all property data either from the tables or from the chart, but not from both.

*Source of Data: American Society of Heating, Refrigerating, and Air-Conditioning Engineers, Inc., Atlanta, GA.*

*(a)* Low pressures, $0 < P_R < 1.0$

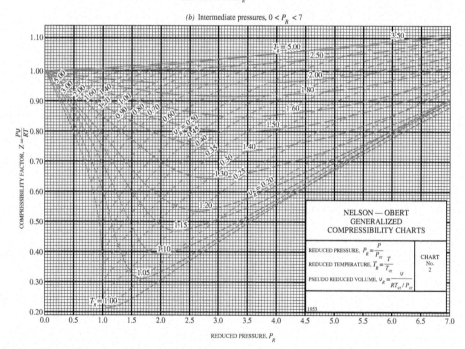

*(b)* Intermediate pressures, $0 < P_R < 7$

## FIGURE A–15

Nelson–Obert generalized compressibility chart.

*Used with permission of Dr. Edward E. Obert, University of Wisconsin.*

## TABLE A–16

Properties of the atmosphere at high altitude

| Altitude, m | Temperature, °C | Pressure, kPa | Gravity $g$, m/s$^2$ | Speed of sound, m/s | Density, kg/m$^3$ | Viscosity $\mu$, kg/m·s | Thermal conductivity, W/m·K |
|---|---|---|---|---|---|---|---|
| 0 | 15.00 | 101.33 | 9.807 | 340.3 | 1.225 | $1.789 \times 10^{-5}$ | 0.0253 |
| 200 | 13.70 | 98.95 | 9.806 | 339.5 | 1.202 | $1.783 \times 10^{-5}$ | 0.0252 |
| 400 | 12.40 | 96.61 | 9.805 | 338.8 | 1.179 | $1.777 \times 10^{-5}$ | 0.0252 |
| 600 | 11.10 | 94.32 | 9.805 | 338.0 | 1.156 | $1.771 \times 10^{-5}$ | 0.0251 |
| 800 | 9.80 | 92.08 | 9.804 | 337.2 | 1.134 | $1.764 \times 10^{-5}$ | 0.0250 |
| 1000 | 8.50 | 89.88 | 9.804 | 336.4 | 1.112 | $1.758 \times 10^{-5}$ | 0.0249 |
| 1200 | 7.20 | 87.72 | 9.803 | 335.7 | 1.090 | $1.752 \times 10^{-5}$ | 0.0248 |
| 1400 | 5.90 | 85.60 | 9.802 | 334.9 | 1.069 | $1.745 \times 10^{-5}$ | 0.0247 |
| 1600 | 4.60 | 83.53 | 9.802 | 334.1 | 1.048 | $1.739 \times 10^{-5}$ | 0.0245 |
| 1800 | 3.30 | 81.49 | 9.801 | 333.3 | 1.027 | $1.732 \times 10^{-5}$ | 0.0244 |
| 2000 | 2.00 | 79.50 | 9.800 | 332.5 | 1.007 | $1.726 \times 10^{-5}$ | 0.0243 |
| 2200 | 0.70 | 77.55 | 9.800 | 331.7 | 0.987 | $1.720 \times 10^{-5}$ | 0.0242 |
| 2400 | −0.59 | 75.63 | 9.799 | 331.0 | 0.967 | $1.713 \times 10^{-5}$ | 0.0241 |
| 2600 | −1.89 | 73.76 | 9.799 | 330.2 | 0.947 | $1.707 \times 10^{-5}$ | 0.0240 |
| 2800 | −3.19 | 71.92 | 9.798 | 329.4 | 0.928 | $1.700 \times 10^{-5}$ | 0.0239 |
| 3000 | −4.49 | 70.12 | 9.797 | 328.6 | 0.909 | $1.694 \times 10^{-5}$ | 0.0238 |
| 3200 | −5.79 | 68.36 | 9.797 | 327.8 | 0.891 | $1.687 \times 10^{-5}$ | 0.0237 |
| 3400 | −7.09 | 66.63 | 9.796 | 327.0 | 0.872 | $1.681 \times 10^{-5}$ | 0.0236 |
| 3600 | −8.39 | 64.94 | 9.796 | 326.2 | 0.854 | $1.674 \times 10^{-5}$ | 0.0235 |
| 3800 | −9.69 | 63.28 | 9.795 | 325.4 | 0.837 | $1.668 \times 10^{-5}$ | 0.0234 |
| 4000 | −10.98 | 61.66 | 9.794 | 324.6 | 0.819 | $1.661 \times 10^{-5}$ | 0.0233 |
| 4200 | −12.3 | 60.07 | 9.794 | 323.8 | 0.802 | $1.655 \times 10^{-5}$ | 0.0232 |
| 4400 | −13.6 | 58.52 | 9.793 | 323.0 | 0.785 | $1.648 \times 10^{-5}$ | 0.0231 |
| 4600 | −14.9 | 57.00 | 9.793 | 322.2 | 0.769 | $1.642 \times 10^{-5}$ | 0.0230 |
| 4800 | −16.2 | 55.51 | 9.792 | 321.4 | 0.752 | $1.635 \times 10^{-5}$ | 0.0229 |
| 5000 | −17.5 | 54.05 | 9.791 | 320.5 | 0.736 | $1.628 \times 10^{-5}$ | 0.0228 |
| 5200 | −18.8 | 52.62 | 9.791 | 319.7 | 0.721 | $1.622 \times 10^{-5}$ | 0.0227 |
| 5400 | −20.1 | 51.23 | 9.790 | 318.9 | 0.705 | $1.615 \times 10^{-5}$ | 0.0226 |
| 5600 | −21.4 | 49.86 | 9.789 | 318.1 | 0.690 | $1.608 \times 10^{-5}$ | 0.0224 |
| 5800 | −22.7 | 48.52 | 9.785 | 317.3 | 0.675 | $1.602 \times 10^{-5}$ | 0.0223 |
| 6000 | −24.0 | 47.22 | 9.788 | 316.5 | 0.660 | $1.595 \times 10^{-5}$ | 0.0222 |
| 6200 | −25.3 | 45.94 | 9.788 | 315.6 | 0.646 | $1.588 \times 10^{-5}$ | 0.0221 |
| 6400 | −26.6 | 44.69 | 9.787 | 314.8 | 0.631 | $1.582 \times 10^{-5}$ | 0.0220 |
| 6600 | −27.9 | 43.47 | 9.786 | 314.0 | 0.617 | $1.575 \times 10^{-5}$ | 0.0219 |
| 6800 | −29.2 | 42.27 | 9.785 | 313.1 | 0.604 | $1.568 \times 10^{-5}$ | 0.0218 |
| 7000 | −30.5 | 41.11 | 9.785 | 312.3 | 0.590 | $1.561 \times 10^{-5}$ | 0.0217 |
| 8000 | −36.9 | 35.65 | 9.782 | 308.1 | 0.526 | $1.527 \times 10^{-5}$ | 0.0212 |
| 9000 | −43.4 | 30.80 | 9.779 | 303.8 | 0.467 | $1.493 \times 10^{-5}$ | 0.0206 |
| 10,000 | −49.9 | 26.50 | 9.776 | 299.5 | 0.414 | $1.458 \times 10^{-5}$ | 0.0201 |
| 12,000 | −56.5 | 19.40 | 9.770 | 295.1 | 0.312 | $1.422 \times 10^{-5}$ | 0.0195 |
| 14,000 | −56.5 | 14.17 | 9.764 | 295.1 | 0.228 | $1.422 \times 10^{-5}$ | 0.0195 |
| 16,000 | −56.5 | 10.53 | 9.758 | 295.1 | 0.166 | $1.422 \times 10^{-5}$ | 0.0195 |
| 18,000 | −56.5 | 7.57 | 9.751 | 295.1 | 0.122 | $1.422 \times 10^{-5}$ | 0.0195 |

*Source of Data:* U.S. Standard Atmosphere Supplements, U.S. Government Printing Office, 1966. Based on year-round mean conditions at 45° latitude and varies with the time of the year and the weather patterns. The conditions at sea level ($z = 0$) are taken to be $P = 101.325$ kPa, $T = 15$°C, $\rho = 1.2250$ kg/m$^3$, $g = 9.80665$ m$^2$/s.

## TABLE A–17

Ideal-gas properties of air

| T K | h kJ/kg | $P_r$ | u kJ/kg | $v_r$ | $s°$ kJ/kg·K | T K | h kJ/kg | $P_r$ | u kJ/kg | $v_r$ | $s°$ kJ/kg·K |
|---|---|---|---|---|---|---|---|---|---|---|---|
| 200 | 199.97 | 0.3363 | 142.56 | 1707.0 | 1.29559 | 580 | 586.04 | 14.38 | 419.55 | 115.7 | 2.37348 |
| 210 | 209.97 | 0.3987 | 149.69 | 1512.0 | 1.34444 | 590 | 596.52 | 15.31 | 427.15 | 110.6 | 2.39140 |
| 220 | 219.97 | 0.4690 | 156.82 | 1346.0 | 1.39105 | 600 | 607.02 | 16.28 | 434.78 | 105.8 | 2.40902 |
| 230 | 230.02 | 0.5477 | 164.00 | 1205.0 | 1.43557 | 610 | 617.53 | 17.30 | 442.42 | 101.2 | 2.42644 |
| 240 | 240.02 | 0.6355 | 171.13 | 1084.0 | 1.47824 | 620 | 628.07 | 18.36 | 450.09 | 96.92 | 2.44356 |
| 250 | 250.05 | 0.7329 | 178.28 | 979.0 | 1.51917 | 630 | 638.63 | 19.84 | 457.78 | 92.84 | 2.46048 |
| 260 | 260.09 | 0.8405 | 185.45 | 887.8 | 1.55848 | 640 | 649.22 | 20.64 | 465.50 | 88.99 | 2.47716 |
| 270 | 270.11 | 0.9590 | 192.60 | 808.0 | 1.59634 | 650 | 659.84 | 21.86 | 473.25 | 85.34 | 2.49364 |
| 280 | 280.13 | 1.0889 | 199.75 | 738.0 | 1.63279 | 660 | 670.47 | 23.13 | 481.01 | 81.89 | 2.50985 |
| 285 | 285.14 | 1.1584 | 203.33 | 706.1 | 1.65055 | 670 | 681.14 | 24.46 | 488.81 | 78.61 | 2.52589 |
| 290 | 290.16 | 1.2311 | 206.91 | 676.1 | 1.66802 | 680 | 691.82 | 25.85 | 496.62 | 75.50 | 2.54175 |
| 295 | 295.17 | 1.3068 | 210.49 | 647.9 | 1.68515 | 690 | 702.52 | 27.29 | 504.45 | 72.56 | 2.55731 |
| 298 | 298.18 | 1.3543 | 212.64 | 631.9 | 1.69528 | 700 | 713.27 | 28.80 | 512.33 | 69.76 | 2.57277 |
| 300 | 300.19 | 1.3860 | 214.07 | 621.2 | 1.70203 | 710 | 724.04 | 30.38 | 520.23 | 67.07 | 2.58810 |
| 305 | 305.22 | 1.4686 | 217.67 | 596.0 | 1.71865 | 720 | 734.82 | 32.02 | 528.14 | 64.53 | 2.60319 |
| 310 | 310.24 | 1.5546 | 221.25 | 572.3 | 1.73498 | 730 | 745.62 | 33.72 | 536.07 | 62.13 | 2.61803 |
| 315 | 315.27 | 1.6442 | 224.85 | 549.8 | 1.75106 | 740 | 756.44 | 35.50 | 544.02 | 59.82 | 2.63280 |
| 320 | 320.29 | 1.7375 | 228.42 | 528.6 | 1.76690 | 750 | 767.29 | 37.35 | 551.99 | 57.63 | 2.64737 |
| 325 | 325.31 | 1.8345 | 232.02 | 508.4 | 1.78249 | 760 | 778.18 | 39.27 | 560.01 | 55.54 | 2.66176 |
| 330 | 330.34 | 1.9352 | 235.61 | 489.4 | 1.79783 | 780 | 800.03 | 43.35 | 576.12 | 51.64 | 2.69013 |
| 340 | 340.42 | 2.149 | 242.82 | 454.1 | 1.82790 | 800 | 821.95 | 47.75 | 592.30 | 48.08 | 2.71787 |
| 350 | 350.49 | 2.379 | 250.02 | 422.2 | 1.85708 | 820 | 843.98 | 52.59 | 608.59 | 44.84 | 2.74504 |
| 360 | 360.58 | 2.626 | 257.24 | 393.4 | 1.88543 | 840 | 866.08 | 57.60 | 624.95 | 41.85 | 2.77170 |
| 370 | 370.67 | 2.892 | 264.46 | 367.2 | 1.91313 | 860 | 888.27 | 63.09 | 641.40 | 39.12 | 2.79783 |
| 380 | 380.77 | 3.176 | 271.69 | 343.4 | 1.94001 | 880 | 910.56 | 68.98 | 657.95 | 36.61 | 2.82344 |
| 390 | 390.88 | 3.481 | 278.93 | 321.5 | 1.96633 | 900 | 932.93 | 75.29 | 674.58 | 34.31 | 2.84856 |
| 400 | 400.98 | 3.806 | 286.16 | 301.6 | 1.99194 | 920 | 955.38 | 82.05 | 691.28 | 32.18 | 2.87324 |
| 410 | 411.12 | 4.153 | 293.43 | 283.3 | 2.01699 | 940 | 977.92 | 89.28 | 708.08 | 30.22 | 2.89748 |
| 420 | 421.26 | 4.522 | 300.69 | 266.6 | 2.04142 | 960 | 1000.55 | 97.00 | 725.02 | 28.40 | 2.92128 |
| 430 | 431.43 | 4.915 | 307.99 | 251.1 | 2.06533 | 980 | 1023.25 | 105.2 | 741.98 | 26.73 | 2.94468 |
| 440 | 441.61 | 5.332 | 315.30 | 236.8 | 2.08870 | 1000 | 1046.04 | 114.0 | 758.94 | 25.17 | 2.96770 |
| 450 | 451.80 | 5.775 | 322.62 | 223.6 | 2.11161 | 1020 | 1068.89 | 123.4 | 776.10 | 23.72 | 2.99034 |
| 460 | 462.02 | 6.245 | 329.97 | 211.4 | 2.13407 | 1040 | 1091.85 | 133.3 | 793.36 | 23.29 | 3.01260 |
| 470 | 472.24 | 6.742 | 337.32 | 200.1 | 2.15604 | 1060 | 1114.86 | 143.9 | 810.62 | 21.14 | 3.03449 |
| 480 | 482.49 | 7.268 | 344.70 | 189.5 | 2.17760 | 1080 | 1137.89 | 155.2 | 827.88 | 19.98 | 3.05608 |
| 490 | 492.74 | 7.824 | 352.08 | 179.7 | 2.19876 | 1100 | 1161.07 | 167.1 | 845.33 | 18.896 | 3.07732 |
| 500 | 503.02 | 8.411 | 359.49 | 170.6 | 2.21952 | 1120 | 1184.28 | 179.7 | 862.79 | 17.886 | 3.09825 |
| 510 | 513.32 | 9.031 | 366.92 | 162.1 | 2.23993 | 1140 | 1207.57 | 193.1 | 880.35 | 16.946 | 3.11883 |
| 520 | 523.63 | 9.684 | 374.36 | 154.1 | 2.25997 | 1160 | 1230.92 | 207.2 | 897.91 | 16.064 | 3.13916 |
| 530 | 533.98 | 10.37 | 381.84 | 146.7 | 2.27967 | 1180 | 1254.34 | 222.2 | 915.57 | 15.241 | 3.15916 |
| 540 | 544.35 | 11.10 | 389.34 | 139.7 | 2.29906 | 1200 | 1277.79 | 238.0 | 933.33 | 14.470 | 3.17888 |
| 550 | 554.74 | 11.86 | 396.86 | 133.1 | 2.31809 | 1220 | 1301.31 | 254.7 | 951.09 | 13.747 | 3.19834 |
| 560 | 565.17 | 12.66 | 404.42 | 127.0 | 2.33685 | 1240 | 1324.93 | 272.3 | 968.95 | 13.069 | 3.21751 |
| 570 | 575.59 | 13.50 | 411.97 | 121.2 | 2.35531 | | | | | | |

## TABLE A–17

Ideal-gas properties of air  (Concluded)

| T K | h kJ/kg | $P_r$ | u kJ/kg | $v_r$ | $s°$ kJ/kg·K | T K | h kJ/kg | $P_r$ | u kJ/kg | $v_r$ | $s°$ kJ/kg·K |
|---|---|---|---|---|---|---|---|---|---|---|---|
| 1260 | 1348.55 | 290.8 | 986.90 | 12.435 | 3.23638 | 1600 | 1757.57 | 791.2 | 1298.30 | 5.804 | 3.52364 |
| 1280 | 1372.24 | 310.4 | 1004.76 | 11.835 | 3.25510 | 1620 | 1782.00 | 834.1 | 1316.96 | 5.574 | 3.53879 |
| 1300 | 1395.97 | 330.9 | 1022.82 | 11.275 | 3.27345 | 1640 | 1806.46 | 878.9 | 1335.72 | 5.355 | 3.55381 |
| 1320 | 1419.76 | 352.5 | 1040.88 | 10.747 | 3.29160 | 1660 | 1830.96 | 925.6 | 1354.48 | 5.147 | 3.56867 |
| 1340 | 1443.60 | 375.3 | 1058.94 | 10.247 | 3.30959 | 1680 | 1855.50 | 974.2 | 1373.24 | 4.949 | 3.58335 |
| 1360 | 1467.49 | 399.1 | 1077.10 | 9.780 | 3.32724 | 1700 | 1880.1 | 1025 | 1392.7 | 4.761 | 3.5979 |
| 1380 | 1491.44 | 424.2 | 1095.26 | 9.337 | 3.34474 | 1750 | 1941.6 | 1161 | 1439.8 | 4.328 | 3.6336 |
| 1400 | 1515.42 | 450.5 | 1113.52 | 8.919 | 3.36200 | 1800 | 2003.3 | 1310 | 1487.2 | 3.994 | 3.6684 |
| 1420 | 1539.44 | 478.0 | 1131.77 | 8.526 | 3.37901 | 1850 | 2065.3 | 1475 | 1534.9 | 3.601 | 3.7023 |
| 1440 | 1563.51 | 506.9 | 1150.13 | 8.153 | 3.39586 | 1900 | 2127.4 | 1655 | 1582.6 | 3.295 | 3.7354 |
| 1460 | 1587.63 | 537.1 | 1168.49 | 7.801 | 3.41247 | 1950 | 2189.7 | 1852 | 1630.6 | 3.022 | 3.7677 |
| 1480 | 1611.79 | 568.8 | 1186.95 | 7.468 | 3.42892 | 2000 | 2252.1 | 2068 | 1678.7 | 2.776 | 3.7994 |
| 1500 | 1635.97 | 601.9 | 1205.41 | 7.152 | 3.44516 | 2050 | 2314.6 | 2303 | 1726.8 | 2.555 | 3.8303 |
| 1520 | 1660.23 | 636.5 | 1223.87 | 6.854 | 3.46120 | 2100 | 2377.7 | 2559 | 1775.3 | 2.356 | 3.8605 |
| 1540 | 1684.51 | 672.8 | 1242.43 | 6.569 | 3.47712 | 2150 | 2440.3 | 2837 | 1823.8 | 2.175 | 3.8901 |
| 1560 | 1708.82 | 710.5 | 1260.99 | 6.301 | 3.49276 | 2200 | 2503.2 | 3138 | 1872.4 | 2.012 | 3.9191 |
| 1580 | 1733.17 | 750.0 | 1279.65 | 6.046 | 3.50829 | 2250 | 2566.4 | 3464 | 1921.3 | 1.864 | 3.9474 |

*Note:* The properties $P_r$ (relative pressure) and $v_r$ (relative specific volume) are dimensionless quantities used in the analysis of isentropic processes, and should not be confused with the properties pressure and specific volume.

*Source of Data:* Kenneth Wark, *Thermodynamics,* 4th ed. (New York: McGraw-Hill, 1983), pp. 785–86, table A–5. Originally published in J. H. Keenan and J. Kaye, *Gas Tables* (New York: John Wiley & Sons, 1948).

## TABLE A–18

Ideal-gas properties of nitrogen, $N_2$

| $T$ K | $\bar{h}$ kJ/kmol | $\bar{u}$ kJ/kmol | $\bar{s}°$ kJ/kmol·K | $T$ K | $\bar{h}$ kJ/kmol | $\bar{u}$ kJ/kmol | $\bar{s}°$ kJ/kmol·K |
|---|---|---|---|---|---|---|---|
| 0 | 0 | 0 | 0 | 600 | 17,563 | 12,574 | 212.066 |
| 220 | 6,391 | 4,562 | 182.639 | 610 | 17,864 | 12,792 | 212.564 |
| 230 | 6,683 | 4,770 | 183.938 | 620 | 18,166 | 13,011 | 213.055 |
| 240 | 6,975 | 4,979 | 185.180 | 630 | 18,468 | 13,230 | 213.541 |
| 250 | 7,266 | 5,188 | 186.370 | 640 | 18,772 | 13,450 | 214.018 |
| 260 | 7,558 | 5,396 | 187.514 | 650 | 19,075 | 13,671 | 214.489 |
| 270 | 7,849 | 5,604 | 188.614 | 660 | 19,380 | 13,892 | 214.954 |
| 280 | 8,141 | 5,813 | 189.673 | 670 | 19,685 | 14,114 | 215.413 |
| 290 | 8,432 | 6,021 | 190.695 | 680 | 19,991 | 14,337 | 215.866 |
| 298 | 8,669 | 6,190 | 191.502 | 690 | 20,297 | 14,560 | 216.314 |
| 300 | 8,723 | 6,229 | 191.682 | 700 | 20,604 | 14,784 | 216.756 |
| 310 | 9,014 | 6,437 | 192.638 | 710 | 20,912 | 15,008 | 217.192 |
| 320 | 9,306 | 6,645 | 193.562 | 720 | 21,220 | 15,234 | 217.624 |
| 330 | 9,597 | 6,853 | 194.459 | 730 | 21,529 | 15,460 | 218.059 |
| 340 | 9,888 | 7,061 | 195.328 | 740 | 21,839 | 15,686 | 218.472 |
| 350 | 10,180 | 7,270 | 196.173 | 750 | 22,149 | 15,913 | 218.889 |
| 360 | 10,471 | 7,478 | 196.995 | 760 | 22,460 | 16,141 | 219.301 |
| 370 | 10,763 | 7,687 | 197.794 | 770 | 22,772 | 16,370 | 219.709 |
| 380 | 11,055 | 7,895 | 198.572 | 780 | 23,085 | 16,599 | 220.113 |
| 390 | 11,347 | 8,104 | 199.331 | 790 | 23,398 | 16,830 | 220.512 |
| 400 | 11,640 | 8,314 | 200.071 | 800 | 23,714 | 17,061 | 220.907 |
| 410 | 11,932 | 8,523 | 200.794 | 810 | 24,027 | 17,292 | 221.298 |
| 420 | 12,225 | 8,733 | 201.499 | 820 | 24,342 | 17,524 | 221.684 |
| 430 | 12,518 | 8,943 | 202.189 | 830 | 24,658 | 17,757 | 222.067 |
| 440 | 12,811 | 9,153 | 202.863 | 840 | 24,974 | 17,990 | 222.447 |
| 450 | 13,105 | 9,363 | 203.523 | 850 | 25,292 | 18,224 | 222.822 |
| 460 | 13,399 | 9,574 | 204.170 | 860 | 25,610 | 18,459 | 223.194 |
| 470 | 13,693 | 9,786 | 204.803 | 870 | 25,928 | 18,695 | 223.562 |
| 480 | 13,988 | 9,997 | 205.424 | 880 | 26,248 | 18,931 | 223.927 |
| 490 | 14,285 | 10,210 | 206.033 | 890 | 26,568 | 19,168 | 224.288 |
| 500 | 14,581 | 10,423 | 206.630 | 900 | 26,890 | 19,407 | 224.647 |
| 510 | 14,876 | 10,635 | 207.216 | 910 | 27,210 | 19,644 | 225.002 |
| 520 | 15,172 | 10,848 | 207.792 | 920 | 27,532 | 19,883 | 225.353 |
| 530 | 15,469 | 11,062 | 208.358 | 930 | 27,854 | 20,122 | 225.701 |
| 540 | 15,766 | 11,277 | 208.914 | 940 | 28,178 | 20,362 | 226.047 |
| 550 | 16,064 | 11,492 | 209.461 | 950 | 28,501 | 20,603 | 226.389 |
| 560 | 16,363 | 11,707 | 209.999 | 960 | 28,826 | 20,844 | 226.728 |
| 570 | 16,662 | 11,923 | 210.528 | 970 | 29,151 | 21,086 | 227.064 |
| 580 | 16,962 | 12,139 | 211.049 | 980 | 29,476 | 21,328 | 227.398 |
| 590 | 17,262 | 12,356 | 211.562 | 990 | 29,803 | 21,571 | 227.728 |

Ideal-gas properties of nitrogen, N$_2$ *(Concluded)*

| $T$ K | $\bar{h}$ kJ/kmol | $\bar{u}$ kJ/kmol | $\bar{s}°$ kJ/kmol·K | $T$ K | $\bar{h}$ kJ/kmol | $\bar{u}$ kJ/kmol | $\bar{s}°$ kJ/kmol·K |
|---|---|---|---|---|---|---|---|
| 1000 | 30,129 | 21,815 | 228.057 | 1760 | 56,227 | 41,594 | 247.396 |
| 1020 | 30,784 | 22,304 | 228.706 | 1780 | 56,938 | 42,139 | 247.798 |
| 1040 | 31,442 | 22,795 | 229.344 | 1800 | 57,651 | 42,685 | 248.195 |
| 1060 | 32,101 | 23,288 | 229.973 | 1820 | 58,363 | 43,231 | 248.589 |
| 1080 | 32,762 | 23,782 | 230.591 | 1840 | 59,075 | 43,777 | 248.979 |
| 1100 | 33,426 | 24,280 | 231.199 | 1860 | 59,790 | 44,324 | 249.365 |
| 1120 | 34,092 | 24,780 | 231.799 | 1880 | 60,504 | 44,873 | 249.748 |
| 1140 | 34,760 | 25,282 | 232.391 | 1900 | 61,220 | 45,423 | 250.128 |
| 1160 | 35,430 | 25,786 | 232.973 | 1920 | 61,936 | 45,973 | 250.502 |
| 1180 | 36,104 | 26,291 | 233.549 | 1940 | 62,654 | 46,524 | 250.874 |
| 1200 | 36,777 | 26,799 | 234.115 | 1960 | 63,381 | 47,075 | 251.242 |
| 1220 | 37,452 | 27,308 | 234.673 | 1980 | 64,090 | 47,627 | 251.607 |
| 1240 | 38,129 | 27,819 | 235.223 | 2000 | 64,810 | 48,181 | 251.969 |
| 1260 | 38,807 | 28,331 | 235.766 | 2050 | 66,612 | 49,567 | 252.858 |
| 1280 | 39,488 | 28,845 | 236.302 | 2100 | 68,417 | 50,957 | 253.726 |
| 1300 | 40,170 | 29,361 | 236.831 | 2150 | 70,226 | 52,351 | 254.578 |
| 1320 | 40,853 | 29,378 | 237.353 | 2200 | 72,040 | 53,749 | 255.412 |
| 1340 | 41,539 | 30,398 | 237.867 | 2250 | 73,856 | 55,149 | 256.227 |
| 1360 | 42,227 | 30,919 | 238.376 | 2300 | 75,676 | 56,553 | 257.027 |
| 1380 | 42,915 | 31,441 | 238.878 | 2350 | 77,496 | 57,958 | 257.810 |
| 1400 | 43,605 | 31,964 | 239.375 | 2400 | 79,320 | 59,366 | 258.580 |
| 1420 | 44,295 | 32,489 | 239.865 | 2450 | 81,149 | 60,779 | 259.332 |
| 1440 | 44,988 | 33,014 | 240.350 | 2500 | 82,981 | 62,195 | 260.073 |
| 1460 | 45,682 | 33,543 | 240.827 | 2550 | 84,814 | 63,613 | 260.799 |
| 1480 | 46,377 | 34,071 | 241.301 | 2600 | 86,650 | 65,033 | 261.512 |
| 1500 | 47,073 | 34,601 | 241.768 | 2650 | 88,488 | 66,455 | 262.213 |
| 1520 | 47,771 | 35,133 | 242.228 | 2700 | 90,328 | 67,880 | 262.902 |
| 1540 | 48,470 | 35,665 | 242.685 | 2750 | 92,171 | 69,306 | 263.577 |
| 1560 | 49,168 | 36,197 | 243.137 | 2800 | 94,014 | 70,734 | 264.241 |
| 1580 | 49,869 | 36,732 | 243.585 | 2850 | 95,859 | 72,163 | 264.895 |
| 1600 | 50,571 | 37,268 | 244.028 | 2900 | 97,705 | 73,593 | 265.538 |
| 1620 | 51,275 | 37,806 | 244.464 | 2950 | 99,556 | 75,028 | 266.170 |
| 1640 | 51,980 | 38,344 | 244.896 | 3000 | 101,407 | 76,464 | 266.793 |
| 1660 | 52,686 | 38,884 | 245.324 | 3050 | 103,260 | 77,902 | 267.404 |
| 1680 | 53,393 | 39,424 | 245.747 | 3100 | 105,115 | 79,341 | 268.007 |
| 1700 | 54,099 | 39,965 | 246.166 | 3150 | 106,972 | 80,782 | 268.601 |
| 1720 | 54,807 | 40,507 | 246.580 | 3200 | 108,830 | 82,224 | 269.186 |
| 1740 | 55,516 | 41,049 | 246.990 | 3250 | 110,690 | 83,668 | 269.763 |

*Source of Data:* Tables A–18 through A–25 are adapted from Kenneth Wark, *Thermodynamics,* 4th ed. (New York: McGraw-Hill, 1983), pp. 787–98. Originally published in JANAF, *Thermochemical Tables,* NSRDS-NBS-37, 1971.

## TABLE A-19

Ideal-gas properties of oxygen, $O_2$

| $T$ K | $\overline{h}$ kJ/kmol | $\overline{u}$ kJ/kmol | $\overline{s}°$ kJ/kmol·K | $T$ K | $\overline{h}$ kJ/kmol | $\overline{u}$ kJ/kmol | $\overline{s}°$ kJ/kmol·K |
|---|---|---|---|---|---|---|---|
| 0 | 0 | 0 | 0 | 600 | 17,929 | 12,940 | 226.346 |
| 220 | 6,404 | 4,575 | 196.171 | 610 | 18,250 | 13,178 | 226.877 |
| 230 | 6,694 | 4,782 | 197.461 | 620 | 18,572 | 13,417 | 227.400 |
| 240 | 6,984 | 4,989 | 198.696 | 630 | 18,895 | 13,657 | 227.918 |
| 250 | 7,275 | 5,197 | 199.885 | 640 | 19,219 | 13,898 | 228.429 |
| 260 | 7,566 | 5,405 | 201.027 | 650 | 19,544 | 14,140 | 228.932 |
| 270 | 7,858 | 5,613 | 202.128 | 660 | 19,870 | 14,383 | 229.430 |
| 280 | 8,150 | 5,822 | 203.191 | 670 | 20,197 | 14,626 | 229.920 |
| 290 | 8,443 | 6,032 | 204.218 | 680 | 20,524 | 14,871 | 230.405 |
| 298 | 8,682 | 6,203 | 205.033 | 690 | 20,854 | 15,116 | 230.885 |
| 300 | 8,736 | 6,242 | 205.213 | 700 | 21,184 | 15,364 | 231.358 |
| 310 | 9,030 | 6,453 | 206.177 | 710 | 21,514 | 15,611 | 231.827 |
| 320 | 9,325 | 6,664 | 207.112 | 720 | 21,845 | 15,859 | 232.291 |
| 330 | 9,620 | 6,877 | 208.020 | 730 | 22,177 | 16,107 | 232.748 |
| 340 | 9,916 | 7,090 | 208.904 | 740 | 22,510 | 16,357 | 233.201 |
| 350 | 10,213 | 7,303 | 209.765 | 750 | 22,844 | 16,607 | 233.649 |
| 360 | 10,511 | 7,518 | 210.604 | 760 | 23,178 | 16,859 | 234.091 |
| 370 | 10,809 | 7,733 | 211.423 | 770 | 23,513 | 17,111 | 234.528 |
| 380 | 11,109 | 7,949 | 212.222 | 780 | 23,850 | 17,364 | 234.960 |
| 390 | 11,409 | 8,166 | 213.002 | 790 | 24,186 | 17,618 | 235.387 |
| 400 | 11,711 | 8,384 | 213.765 | 800 | 24,523 | 17,872 | 235.810 |
| 410 | 12,012 | 8,603 | 214.510 | 810 | 24,861 | 18,126 | 236.230 |
| 420 | 12,314 | 8,822 | 215.241 | 820 | 25,199 | 18,382 | 236.644 |
| 430 | 12,618 | 9,043 | 215.955 | 830 | 25,537 | 18,637 | 237.055 |
| 440 | 12,923 | 9,264 | 216.656 | 840 | 25,877 | 18,893 | 237.462 |
| 450 | 13,228 | 9,487 | 217.342 | 850 | 26,218 | 19,150 | 237.864 |
| 460 | 13,525 | 9,710 | 218.016 | 860 | 26,559 | 19,408 | 238.264 |
| 470 | 13,842 | 9,935 | 218.676 | 870 | 26,899 | 19,666 | 238.660 |
| 480 | 14,151 | 10,160 | 219.326 | 880 | 27,242 | 19,925 | 239.051 |
| 490 | 14,460 | 10,386 | 219.963 | 890 | 27,584 | 20,185 | 239.439 |
| 500 | 14,770 | 10,614 | 220.589 | 900 | 27,928 | 20,445 | 239.823 |
| 510 | 15,082 | 10,842 | 221.206 | 910 | 28,272 | 20,706 | 240.203 |
| 520 | 15,395 | 11,071 | 221.812 | 920 | 28,616 | 20,967 | 240.580 |
| 530 | 15,708 | 11,301 | 222.409 | 930 | 28,960 | 21,228 | 240.953 |
| 540 | 16,022 | 11,533 | 222.997 | 940 | 29,306 | 21,491 | 241.323 |
| 550 | 16,338 | 11,765 | 223.576 | 950 | 29,652 | 21,754 | 241.689 |
| 560 | 16,654 | 11,998 | 224.146 | 960 | 29,999 | 22,017 | 242.052 |
| 570 | 16,971 | 12,232 | 224.708 | 970 | 30,345 | 22,280 | 242.411 |
| 580 | 17,290 | 12,467 | 225.262 | 980 | 30,692 | 22,544 | 242.768 |
| 590 | 17,609 | 12,703 | 225.808 | 990 | 31,041 | 22,809 | 242.120 |

## TABLE A–19

Ideal-gas properties of oxygen, $O_2$ *(Concluded)*

| $T$ K | $\bar{h}$ kJ/kmol | $\bar{u}$ kJ/kmol | $\bar{s}°$ kJ/kmol·K | $T$ K | $\bar{h}$ kJ/kmol | $\bar{u}$ kJ/kmol | $\bar{s}°$ kJ/kmol·K |
|---|---|---|---|---|---|---|---|
| 1000 | 31,389 | 23,075 | 243.471 | 1760 | 58,880 | 44,247 | 263.861 |
| 1020 | 32,088 | 23,607 | 244.164 | 1780 | 59,624 | 44,825 | 264.283 |
| 1040 | 32,789 | 24,142 | 244.844 | 1800 | 60,371 | 45,405 | 264.701 |
| 1060 | 33,490 | 24,677 | 245.513 | 1820 | 61,118 | 45,986 | 265.113 |
| 1080 | 34,194 | 25,214 | 246.171 | 1840 | 61,866 | 46,568 | 265.521 |
| 1100 | 34,899 | 25,753 | 246.818 | 1860 | 62,616 | 47,151 | 265.925 |
| 1120 | 35,606 | 26,294 | 247.454 | 1880 | 63,365 | 47,734 | 266.326 |
| 1140 | 36,314 | 26,836 | 248.081 | 1900 | 64,116 | 48,319 | 266.722 |
| 1160 | 37,023 | 27,379 | 248.698 | 1920 | 64,868 | 48,904 | 267.115 |
| 1180 | 37,734 | 27,923 | 249.307 | 1940 | 65,620 | 49,490 | 267.505 |
| 1200 | 38,447 | 28,469 | 249.906 | 1960 | 66,374 | 50,078 | 267.891 |
| 1220 | 39,162 | 29,018 | 250.497 | 1980 | 67,127 | 50,665 | 268.275 |
| 1240 | 39,877 | 29,568 | 251.079 | 2000 | 67,881 | 51,253 | 268.655 |
| 1260 | 40,594 | 30,118 | 251.653 | 2050 | 69,772 | 52,727 | 269.588 |
| 1280 | 41,312 | 30,670 | 252.219 | 2100 | 71,668 | 54,208 | 270.504 |
| 1300 | 42,033 | 31,224 | 252.776 | 2150 | 73,573 | 55,697 | 271.399 |
| 1320 | 42,753 | 31,778 | 253.325 | 2200 | 75,484 | 57,192 | 272.278 |
| 1340 | 43,475 | 32,334 | 253.868 | 2250 | 77,397 | 58,690 | 273.136 |
| 1360 | 44,198 | 32,891 | 254.404 | 2300 | 79,316 | 60,193 | 273.891 |
| 1380 | 44,923 | 33,449 | 254.932 | 2350 | 81,243 | 61,704 | 274.809 |
| 1400 | 45,648 | 34,008 | 255.454 | 2400 | 83,174 | 63,219 | 275.625 |
| 1420 | 46,374 | 34,567 | 255.968 | 2450 | 85,112 | 64,742 | 276.424 |
| 1440 | 47,102 | 35,129 | 256.475 | 2500 | 87,057 | 66,271 | 277.207 |
| 1460 | 47,831 | 35,692 | 256.978 | 2550 | 89,004 | 67,802 | 277.979 |
| 1480 | 48,561 | 36,256 | 257.474 | 2600 | 90,956 | 69,339 | 278.738 |
| 1500 | 49,292 | 36,821 | 257.965 | 2650 | 92,916 | 70,883 | 279.485 |
| 1520 | 50,024 | 37,387 | 258.450 | 2700 | 94,881 | 72,433 | 280.219 |
| 1540 | 50,756 | 37,952 | 258.928 | 2750 | 96,852 | 73,987 | 280.942 |
| 1560 | 51,490 | 38,520 | 259.402 | 2800 | 98,826 | 75,546 | 281.654 |
| 1580 | 52,224 | 39,088 | 259.870 | 2850 | 100,808 | 77,112 | 282.357 |
| 1600 | 52,961 | 39,658 | 260.333 | 2900 | 102,793 | 78,682 | 283.048 |
| 1620 | 53,696 | 40,227 | 260.791 | 2950 | 104,785 | 80,258 | 283.728 |
| 1640 | 54,434 | 40,799 | 261.242 | 3000 | 106,780 | 81,837 | 284.399 |
| 1660 | 55,172 | 41,370 | 261.690 | 3050 | 108,778 | 83,419 | 285.060 |
| 1680 | 55,912 | 41,944 | 262.132 | 3100 | 110,784 | 85,009 | 285.713 |
| 1700 | 56,652 | 42,517 | 262.571 | 3150 | 112,795 | 86,601 | 286.355 |
| 1720 | 57,394 | 43,093 | 263.005 | 3200 | 114,809 | 88,203 | 286.989 |
| 1740 | 58,136 | 43,669 | 263.435 | 3250 | 116,827 | 89,804 | 287.614 |

Ideal-gas properties of carbon dioxide, $CO_2$

| $T$ K | $\bar{h}$ kJ/kmol | $\bar{u}$ kJ/kmol | $\bar{s}°$ kJ/kmol·K | $T$ K | $\bar{h}$ kJ/kmol | $\bar{u}$ kJ/kmol | $\bar{s}°$ kJ/kmol·K |
|---|---|---|---|---|---|---|---|
| 0 | 0 | 0 | 0 | 600 | 22,280 | 17,291 | 243.199 |
| 220 | 6,601 | 4,772 | 202.966 | 610 | 22,754 | 17,683 | 243.983 |
| 230 | 6,938 | 5,026 | 204.464 | 620 | 23,231 | 18,076 | 244.758 |
| 240 | 7,280 | 5,285 | 205.920 | 630 | 23,709 | 18,471 | 245.524 |
| 250 | 7,627 | 5,548 | 207.337 | 640 | 24,190 | 18,869 | 246.282 |
| 260 | 7,979 | 5,817 | 208.717 | 650 | 24,674 | 19,270 | 247.032 |
| 270 | 8,335 | 6,091 | 210.062 | 660 | 25,160 | 19,672 | 247.773 |
| 280 | 8,697 | 6,369 | 211.376 | 670 | 25,648 | 20,078 | 248.507 |
| 290 | 9,063 | 6,651 | 212.660 | 680 | 26,138 | 20,484 | 249.233 |
| 298 | 9,364 | 6,885 | 213.685 | 690 | 26,631 | 20,894 | 249.952 |
| 300 | 9,431 | 6,939 | 213.915 | 700 | 27,125 | 21,305 | 250.663 |
| 310 | 9,807 | 7,230 | 215.146 | 710 | 27,622 | 21,719 | 251.368 |
| 320 | 10,186 | 7,526 | 216.351 | 720 | 28,121 | 22,134 | 252.065 |
| 330 | 10,570 | 7,826 | 217.534 | 730 | 28,622 | 22,522 | 252.755 |
| 340 | 10,959 | 8,131 | 218.694 | 740 | 29,124 | 22,972 | 253.439 |
| 350 | 11,351 | 8,439 | 219.831 | 750 | 29,629 | 23,393 | 254.117 |
| 360 | 11,748 | 8,752 | 220.948 | 760 | 30,135 | 23,817 | 254.787 |
| 370 | 12,148 | 9,068 | 222.044 | 770 | 30,644 | 24,242 | 255.452 |
| 380 | 12,552 | 9,392 | 223.122 | 780 | 31,154 | 24,669 | 256.110 |
| 390 | 12,960 | 9,718 | 224.182 | 790 | 31,665 | 25,097 | 256.762 |
| 400 | 13,372 | 10,046 | 225.225 | 800 | 32,179 | 25,527 | 257.408 |
| 410 | 13,787 | 10,378 | 226.250 | 810 | 32,694 | 25,959 | 258.048 |
| 420 | 14,206 | 10,714 | 227.258 | 820 | 33,212 | 26,394 | 258.682 |
| 430 | 14,628 | 11,053 | 228.252 | 830 | 33,730 | 26,829 | 259.311 |
| 440 | 15,054 | 11,393 | 229.230 | 840 | 34,251 | 27,267 | 259.934 |
| 450 | 15,483 | 11,742 | 230.194 | 850 | 34,773 | 27,706 | 260.551 |
| 460 | 15,916 | 12,091 | 231.144 | 860 | 35,296 | 28,125 | 261.164 |
| 470 | 16,351 | 12,444 | 232.080 | 870 | 35,821 | 28,588 | 261.770 |
| 480 | 16,791 | 12,800 | 233.004 | 880 | 36,347 | 29,031 | 262.371 |
| 490 | 17,232 | 13,158 | 233.916 | 890 | 36,876 | 29,476 | 262.968 |
| 500 | 17,678 | 13,521 | 234.814 | 900 | 37,405 | 29,922 | 263.559 |
| 510 | 18,126 | 13,885 | 235.700 | 910 | 37,935 | 30,369 | 264.146 |
| 520 | 18,576 | 14,253 | 236.575 | 920 | 38,467 | 30,818 | 264.728 |
| 530 | 19,029 | 14,622 | 237.439 | 930 | 39,000 | 31,268 | 265.304 |
| 540 | 19,485 | 14,996 | 238.292 | 940 | 39,535 | 31,719 | 265.877 |
| 550 | 19,945 | 15,372 | 239.135 | 950 | 40,070 | 32,171 | 266.444 |
| 560 | 20,407 | 15,751 | 239.962 | 960 | 40,607 | 32,625 | 267.007 |
| 570 | 20,870 | 16,131 | 240.789 | 970 | 41,145 | 33,081 | 267.566 |
| 580 | 21,337 | 16,515 | 241.602 | 980 | 41,685 | 33,537 | 268.119 |
| 590 | 21,807 | 16,902 | 242.405 | 990 | 42,226 | 33,995 | 268.670 |

Ideal-gas properties of carbon dioxide, $CO_2$ *(Concluded)*

| $T$ K | $\bar{h}$ kJ/kmol | $\bar{u}$ kJ/kmol | $\bar{s}°$ kJ/kmol·K | $T$ K | $\bar{h}$ kJ/kmol | $\bar{u}$ kJ/kmol | $\bar{s}°$ kJ/kmol·K |
|---|---|---|---|---|---|---|---|
| 1000 | 42,769 | 34,455 | 269.215 | 1760 | 86,420 | 71,787 | 301.543 |
| 1020 | 43,859 | 35,378 | 270.293 | 1780 | 87,612 | 72,812 | 302.217 |
| 1040 | 44,953 | 36,306 | 271.354 | 1800 | 88,806 | 73,840 | 302.884 |
| 1060 | 46,051 | 37,238 | 272.400 | 1820 | 90,000 | 74,868 | 303.544 |
| 1080 | 47,153 | 38,174 | 273.430 | 1840 | 91,196 | 75,897 | 304.198 |
| 1100 | 48,258 | 39,112 | 274.445 | 1860 | 92,394 | 76,929 | 304.845 |
| 1120 | 49,369 | 40,057 | 275.444 | 1880 | 93,593 | 77,962 | 305.487 |
| 1140 | 50,484 | 41,006 | 276.430 | 1900 | 94,793 | 78,996 | 306.122 |
| 1160 | 51,602 | 41,957 | 277.403 | 1920 | 95,995 | 80,031 | 306.751 |
| 1180 | 52,724 | 42,913 | 278.361 | 1940 | 97,197 | 81,067 | 307.374 |
| 1200 | 53,848 | 43,871 | 297.307 | 1960 | 98,401 | 82,105 | 307.992 |
| 1220 | 54,977 | 44,834 | 280.238 | 1980 | 99,606 | 83,144 | 308.604 |
| 1240 | 56,108 | 45,799 | 281.158 | 2000 | 100,804 | 84,185 | 309.210 |
| 1260 | 57,244 | 46,768 | 282.066 | 2050 | 103,835 | 86,791 | 310.701 |
| 1280 | 58,381 | 47,739 | 282.962 | 2100 | 106,864 | 89,404 | 312.160 |
| 1300 | 59,522 | 48,713 | 283.847 | 2150 | 109,898 | 92,023 | 313.589 |
| 1320 | 60,666 | 49,691 | 284.722 | 2200 | 112,939 | 94,648 | 314.988 |
| 1340 | 61,813 | 50,672 | 285.586 | 2250 | 115,984 | 97,277 | 316.356 |
| 1360 | 62,963 | 51,656 | 286.439 | 2300 | 119,035 | 99,912 | 317.695 |
| 1380 | 64,116 | 52,643 | 287.283 | 2350 | 122,091 | 102,552 | 319.011 |
| 1400 | 65,271 | 53,631 | 288.106 | 2400 | 125,152 | 105,197 | 320.302 |
| 1420 | 66,427 | 54,621 | 288.934 | 2450 | 128,219 | 107,849 | 321.566 |
| 1440 | 67,586 | 55,614 | 289.743 | 2500 | 131,290 | 110,504 | 322.808 |
| 1460 | 68,748 | 56,609 | 290.542 | 2550 | 134,368 | 113,166 | 324.026 |
| 1480 | 66,911 | 57,606 | 291.333 | 2600 | 137,449 | 115,832 | 325.222 |
| 1500 | 71,078 | 58,606 | 292.114 | 2650 | 140,533 | 118,500 | 326.396 |
| 1520 | 72,246 | 59,609 | 292.888 | 2700 | 143,620 | 121,172 | 327.549 |
| 1540 | 73,417 | 60,613 | 292.654 | 2750 | 146,713 | 123,849 | 328.684 |
| 1560 | 74,590 | 61,620 | 294.411 | 2800 | 149,808 | 126,528 | 329.800 |
| 1580 | 76,767 | 62,630 | 295.161 | 2850 | 152,908 | 129,212 | 330.896 |
| 1600 | 76,944 | 63,741 | 295.901 | 2900 | 156,009 | 131,898 | 331.975 |
| 1620 | 78,123 | 64,653 | 296.632 | 2950 | 159,117 | 134,589 | 333.037 |
| 1640 | 79,303 | 65,668 | 297.356 | 3000 | 162,226 | 137,283 | 334.084 |
| 1660 | 80,486 | 66,592 | 298.072 | 3050 | 165,341 | 139,982 | 335.114 |
| 1680 | 81,670 | 67,702 | 298.781 | 3100 | 168,456 | 142,681 | 336.126 |
| 1700 | 82,856 | 68,721 | 299.482 | 3150 | 171,576 | 145,385 | 337.124 |
| 1720 | 84,043 | 69,742 | 300.177 | 3200 | 174,695 | 148,089 | 338.109 |
| 1740 | 85,231 | 70,764 | 300.863 | 3250 | 177,822 | 150,801 | 339.069 |

## TABLE A–21

Ideal-gas properties of carbon monoxide, CO

| $T$ K | $\bar{h}$ kJ/kmol | $\bar{u}$ kJ/kmol | $\bar{s}°$ kJ/kmol·K | $T$ K | $\bar{h}$ kJ/kmol | $\bar{u}$ kJ/kmol | $\bar{s}°$ kJ/kmol·K |
|---|---|---|---|---|---|---|---|
| 0 | 0 | 0 | 0 | 600 | 17,611 | 12,622 | 218.204 |
| 220 | 6,391 | 4,562 | 188.683 | 610 | 17,915 | 12,843 | 218.708 |
| 230 | 6,683 | 4,771 | 189.980 | 620 | 18,221 | 13,066 | 219.205 |
| 240 | 6,975 | 4,979 | 191.221 | 630 | 18,527 | 13,289 | 219.695 |
| 250 | 7,266 | 5,188 | 192.411 | 640 | 18,833 | 13,512 | 220.179 |
| 260 | 7,558 | 5,396 | 193.554 | 650 | 19,141 | 13,736 | 220.656 |
| 270 | 7,849 | 5,604 | 194.654 | 660 | 19,449 | 13,962 | 221.127 |
| 280 | 8,140 | 5,812 | 195.713 | 670 | 19,758 | 14,187 | 221.592 |
| 290 | 8,432 | 6,020 | 196.735 | 680 | 20,068 | 14,414 | 222.052 |
| 298 | 8,669 | 6,190 | 197.543 | 690 | 20,378 | 14,641 | 222.505 |
| 300 | 8,723 | 6,229 | 197.723 | 700 | 20,690 | 14,870 | 222.953 |
| 310 | 9,014 | 6,437 | 198.678 | 710 | 21,002 | 15,099 | 223.396 |
| 320 | 9,306 | 6,645 | 199.603 | 720 | 21,315 | 15,328 | 223.833 |
| 330 | 9,597 | 6,854 | 200.500 | 730 | 21,628 | 15,558 | 224.265 |
| 340 | 9,889 | 7,062 | 201.371 | 740 | 21,943 | 15,789 | 224.692 |
| 350 | 10,181 | 7,271 | 202.217 | 750 | 22,258 | 16,022 | 225.115 |
| 360 | 10,473 | 7,480 | 203.040 | 760 | 22,573 | 16,255 | 225.533 |
| 370 | 10,765 | 7,689 | 203.842 | 770 | 22,890 | 16,488 | 225.947 |
| 380 | 11,058 | 7,899 | 204.622 | 780 | 23,208 | 16,723 | 226.357 |
| 390 | 11,351 | 8,108 | 205.383 | 790 | 23,526 | 16,957 | 226.762 |
| 400 | 11,644 | 8,319 | 206.125 | 800 | 23,844 | 17,193 | 227.162 |
| 410 | 11,938 | 8,529 | 206.850 | 810 | 24,164 | 17,429 | 227.559 |
| 420 | 12,232 | 8,740 | 207.549 | 820 | 24,483 | 17,665 | 227.952 |
| 430 | 12,526 | 8,951 | 208.252 | 830 | 24,803 | 17,902 | 228.339 |
| 440 | 12,821 | 9,163 | 208.929 | 840 | 25,124 | 18,140 | 228.724 |
| 450 | 13,116 | 9,375 | 209.593 | 850 | 25,446 | 18,379 | 229.106 |
| 460 | 13,412 | 9,587 | 210.243 | 860 | 25,768 | 18,617 | 229.482 |
| 470 | 13,708 | 9,800 | 210.880 | 870 | 26,091 | 18,858 | 229.856 |
| 480 | 14,005 | 10,014 | 211.504 | 880 | 26,415 | 19,099 | 230.227 |
| 490 | 14,302 | 10,228 | 212.117 | 890 | 26,740 | 19,341 | 230.593 |
| 500 | 14,600 | 10,443 | 212.719 | 900 | 27,066 | 19,583 | 230.957 |
| 510 | 14,898 | 10,658 | 213.310 | 910 | 27,392 | 19,826 | 231.317 |
| 520 | 15,197 | 10,874 | 213.890 | 920 | 27,719 | 20,070 | 231.674 |
| 530 | 15,497 | 11,090 | 214.460 | 930 | 28,046 | 20,314 | 232.028 |
| 540 | 15,797 | 11,307 | 215.020 | 940 | 28,375 | 20,559 | 232.379 |
| 550 | 16,097 | 11,524 | 215.572 | 950 | 28,703 | 20,805 | 232.727 |
| 560 | 16,399 | 11,743 | 216.115 | 960 | 29,033 | 21,051 | 233.072 |
| 570 | 16,701 | 11,961 | 216.649 | 970 | 29,362 | 21,298 | 233.413 |
| 580 | 17,003 | 12,181 | 217.175 | 980 | 29,693 | 21,545 | 233.752 |
| 590 | 17,307 | 12,401 | 217.693 | 990 | 30,024 | 21,793 | 234.088 |

Ideal-gas properties of carbon monoxide, CO *(Concluded)*

| $T$ K | $\bar{h}$ kJ/kmol | $\bar{u}$ kJ/kmol | $\bar{s}°$ kJ/kmol·K | $T$ K | $\bar{h}$ kJ/kmol | $\bar{u}$ kJ/kmol | $\bar{s}°$ kJ/kmol·K |
|---|---|---|---|---|---|---|---|
| 1000 | 30,355 | 22,041 | 234.421 | 1760 | 56,756 | 42,123 | 253.991 |
| 1020 | 31,020 | 22,540 | 235.079 | 1780 | 57,473 | 42,673 | 254.398 |
| 1040 | 31,688 | 23,041 | 235.728 | 1800 | 58,191 | 43,225 | 254.797 |
| 1060 | 32,357 | 23,544 | 236.364 | 1820 | 58,910 | 43,778 | 255.194 |
| 1080 | 33,029 | 24,049 | 236.992 | 1840 | 59,629 | 44,331 | 255.587 |
| 1100 | 33,702 | 24,557 | 237.609 | 1860 | 60,351 | 44,886 | 255.976 |
| 1120 | 34,377 | 25,065 | 238.217 | 1880 | 61,072 | 45,441 | 256.361 |
| 1140 | 35,054 | 25,575 | 238.817 | 1900 | 61,794 | 45,997 | 256.743 |
| 1160 | 35,733 | 26,088 | 239.407 | 1920 | 62,516 | 46,552 | 257.122 |
| 1180 | 36,406 | 26,602 | 239.989 | 1940 | 63,238 | 47,108 | 257.497 |
| 1200 | 37,095 | 27,118 | 240.663 | 1960 | 63,961 | 47,665 | 257.868 |
| 1220 | 37,780 | 27,637 | 241.128 | 1980 | 64,684 | 48,221 | 258.236 |
| 1240 | 38,466 | 28,426 | 241.686 | 2000 | 65,408 | 48,780 | 258.600 |
| 1260 | 39,154 | 28,678 | 242.236 | 2050 | 67,224 | 50,179 | 259.494 |
| 1280 | 39,844 | 29,201 | 242.780 | 2100 | 69,044 | 51,584 | 260.370 |
| 1300 | 40,534 | 29,725 | 243.316 | 2150 | 70,864 | 52,988 | 261.226 |
| 1320 | 41,226 | 30,251 | 243.844 | 2200 | 72,688 | 54,396 | 262.065 |
| 1340 | 41,919 | 30,778 | 244.366 | 2250 | 74,516 | 55,809 | 262.887 |
| 1360 | 42,613 | 31,306 | 244.880 | 2300 | 76,345 | 57,222 | 263.692 |
| 1380 | 43,309 | 31,836 | 245.388 | 2350 | 78,178 | 58,640 | 264.480 |
| 1400 | 44,007 | 32,367 | 245.889 | 2400 | 80,015 | 60,060 | 265.253 |
| 1420 | 44,707 | 32,900 | 246.385 | 2450 | 81,852 | 61,482 | 266.012 |
| 1440 | 45,408 | 33,434 | 246.876 | 2500 | 83,692 | 62,906 | 266.755 |
| 1460 | 46,110 | 33,971 | 247.360 | 2550 | 85,537 | 64,335 | 267.485 |
| 1480 | 46,813 | 34,508 | 247.839 | 2600 | 87,383 | 65,766 | 268.202 |
| 1500 | 47,517 | 35,046 | 248.312 | 2650 | 89,230 | 67,197 | 268.905 |
| 1520 | 48,222 | 35,584 | 248.778 | 2700 | 91,077 | 68,628 | 269.596 |
| 1540 | 48,928 | 36,124 | 249.240 | 2750 | 92,930 | 70,066 | 270.285 |
| 1560 | 49,635 | 36,665 | 249.695 | 2800 | 94,784 | 71,504 | 270.943 |
| 1580 | 50,344 | 37,207 | 250.147 | 2850 | 96,639 | 72,945 | 271.602 |
| 1600 | 51,053 | 37,750 | 250.592 | 2900 | 98,495 | 74,383 | 272.249 |
| 1620 | 51,763 | 38,293 | 251.033 | 2950 | 100,352 | 75,825 | 272.884 |
| 1640 | 52,472 | 38,837 | 251.470 | 3000 | 102,210 | 77,267 | 273.508 |
| 1660 | 53,184 | 39,382 | 251.901 | 3050 | 104,073 | 78,715 | 274.123 |
| 1680 | 53,895 | 39,927 | 252.329 | 3100 | 105,939 | 80,164 | 274.730 |
| 1700 | 54,609 | 40,474 | 252.751 | 3150 | 107,802 | 81,612 | 275.326 |
| 1720 | 55,323 | 41,023 | 253.169 | 3200 | 109,667 | 83,061 | 275.914 |
| 1740 | 56,039 | 41,572 | 253.582 | 3250 | 111,534 | 84,513 | 276.494 |

## TABLE A–22

Ideal-gas properties of hydrogen, $H_2$

| $T$ K | $\bar{h}$ kJ/kmol | $\bar{u}$ kJ/kmol | $\bar{s}°$ kJ/kmol·K | $T$ K | $\bar{h}$ kJ/kmol | $\bar{u}$ kJ/kmol | $\bar{s}°$ kJ/kmol·K |
|---|---|---|---|---|---|---|---|
| 0 | 0 | 0 | 0 | 1440 | 42,808 | 30,835 | 177.410 |
| 260 | 7,370 | 5,209 | 126.636 | 1480 | 44,091 | 31,786 | 178.291 |
| 270 | 7,657 | 5,412 | 127.719 | 1520 | 45,384 | 32,746 | 179.153 |
| 280 | 7,945 | 5,617 | 128.765 | 1560 | 46,683 | 33,713 | 179.995 |
| 290 | 8,233 | 5,822 | 129.775 | 1600 | 47,990 | 34,687 | 180.820 |
| 298 | 8,468 | 5,989 | 130.574 | 1640 | 49,303 | 35,668 | 181.632 |
| 300 | 8,522 | 6,027 | 130.754 | 1680 | 50,622 | 36,654 | 182.428 |
| 320 | 9,100 | 6,440 | 132.621 | 1720 | 51,947 | 37,646 | 183.208 |
| 340 | 9,680 | 6,853 | 134.378 | 1760 | 53,279 | 38,645 | 183.973 |
| 360 | 10,262 | 7,268 | 136.039 | 1800 | 54,618 | 39,652 | 184.724 |
| 380 | 10,843 | 7,684 | 137.612 | 1840 | 55,962 | 40,663 | 185.463 |
| 400 | 11,426 | 8,100 | 139.106 | 1880 | 57,311 | 41,680 | 186.190 |
| 420 | 12,010 | 8,518 | 140.529 | 1920 | 58,668 | 42,705 | 186.904 |
| 440 | 12,594 | 8,936 | 141.888 | 1960 | 60,031 | 43,735 | 187.607 |
| 460 | 13,179 | 9,355 | 143.187 | 2000 | 61,400 | 44,771 | 188.297 |
| 480 | 13,764 | 9,773 | 144.432 | 2050 | 63,119 | 46,074 | 189.148 |
| 500 | 14,350 | 10,193 | 145.628 | 2100 | 64,847 | 47,386 | 189.979 |
| 520 | 14,935 | 10,611 | 146.775 | 2150 | 66,584 | 48,708 | 190.796 |
| 560 | 16,107 | 11,451 | 148.945 | 2200 | 68,328 | 50,037 | 191.598 |
| 600 | 17,280 | 12,291 | 150.968 | 2250 | 70,080 | 51,373 | 192.385 |
| 640 | 18,453 | 13,133 | 152.863 | 2300 | 71,839 | 52,716 | 193.159 |
| 680 | 19,630 | 13,976 | 154.645 | 2350 | 73,608 | 54,069 | 193.921 |
| 720 | 20,807 | 14,821 | 156.328 | 2400 | 75,383 | 55,429 | 194.669 |
| 760 | 21,988 | 15,669 | 157.923 | 2450 | 77,168 | 56,798 | 195.403 |
| 800 | 23,171 | 16,520 | 159.440 | 2500 | 78,960 | 58,175 | 196.125 |
| 840 | 24,359 | 17,375 | 160.891 | 2550 | 80,755 | 59,554 | 196.837 |
| 880 | 25,551 | 18,235 | 162.277 | 2600 | 82,558 | 60,941 | 197.539 |
| 920 | 26,747 | 19,098 | 163.607 | 2650 | 84,368 | 62,335 | 198.229 |
| 960 | 27,948 | 19,966 | 164.884 | 2700 | 86,186 | 63,737 | 198.907 |
| 1000 | 29,154 | 20,839 | 166.114 | 2750 | 88,008 | 65,144 | 199.575 |
| 1040 | 30,364 | 21,717 | 167.300 | 2800 | 89,838 | 66,558 | 200.234 |
| 1080 | 31,580 | 22,601 | 168.449 | 2850 | 91,671 | 67,976 | 200.885 |
| 1120 | 32,802 | 23,490 | 169.560 | 2900 | 93,512 | 69,401 | 201.527 |
| 1160 | 34,028 | 24,384 | 170.636 | 2950 | 95,358 | 70,831 | 202.157 |
| 1200 | 35,262 | 25,284 | 171.682 | 3000 | 97,211 | 72,268 | 202.778 |
| 1240 | 36,502 | 26,192 | 172.698 | 3050 | 99,065 | 73,707 | 203.391 |
| 1280 | 37,749 | 27,106 | 173.687 | 3100 | 100,926 | 75,152 | 203.995 |
| 1320 | 39,002 | 28,027 | 174.652 | 3150 | 102,793 | 76,604 | 204.592 |
| 1360 | 40,263 | 28,955 | 175.593 | 3200 | 104,667 | 78,061 | 205.181 |
| 1400 | 41,530 | 29,889 | 176.510 | 3250 | 106,545 | 79,523 | 205.765 |

## TABLE A–23

Ideal-gas properties of water vapor, $H_2O$

| $T$ K | $\bar{h}$ kJ/kmol | $\bar{u}$ kJ/kmol | $\bar{s}°$ kJ/kmol·K | $T$ K | $\bar{h}$ kJ/kmol | $\bar{u}$ kJ/kmol | $\bar{s}°$ kJ/kmol·K |
|---|---|---|---|---|---|---|---|
| 0 | 0 | 0 | 0 | 600 | 20,402 | 15,413 | 212.920 |
| 220 | 7,295 | 5,466 | 178.576 | 610 | 20,765 | 15,693 | 213.529 |
| 230 | 7,628 | 5,715 | 180.054 | 620 | 21,130 | 15,975 | 214.122 |
| 240 | 7,961 | 5,965 | 181.471 | 630 | 21,495 | 16,257 | 214.707 |
| 250 | 8,294 | 6,215 | 182.831 | 640 | 21,862 | 16,541 | 215.285 |
| 260 | 8,627 | 6,466 | 184.139 | 650 | 22,230 | 16,826 | 215.856 |
| 270 | 8,961 | 6,716 | 185.399 | 660 | 22,600 | 17,112 | 216.419 |
| 280 | 9,296 | 6,968 | 186.616 | 670 | 22,970 | 17,399 | 216.976 |
| 290 | 9,631 | 7,219 | 187.791 | 680 | 23,342 | 17,688 | 217.527 |
| 298 | 9,904 | 7,425 | 188.720 | 690 | 23,714 | 17,978 | 218.071 |
| 300 | 9,966 | 7,472 | 188.928 | 700 | 24,088 | 18,268 | 218.610 |
| 310 | 10,302 | 7,725 | 190.030 | 710 | 24,464 | 18,561 | 219.142 |
| 320 | 10,639 | 7,978 | 191.098 | 720 | 24,840 | 18,854 | 219.668 |
| 330 | 10,976 | 8,232 | 192.136 | 730 | 25,218 | 19,148 | 220.189 |
| 340 | 11,314 | 8,487 | 193.144 | 740 | 25,597 | 19,444 | 220.707 |
| 350 | 11,652 | 8,742 | 194.125 | 750 | 25,977 | 19,741 | 221.215 |
| 360 | 11,992 | 8,998 | 195.081 | 760 | 26,358 | 20,039 | 221.720 |
| 370 | 12,331 | 9,255 | 196.012 | 770 | 26,741 | 20,339 | 222.221 |
| 380 | 12,672 | 9,513 | 196.920 | 780 | 27,125 | 20,639 | 222.717 |
| 390 | 13,014 | 9,771 | 197.807 | 790 | 27,510 | 20,941 | 223.207 |
| 400 | 13,356 | 10,030 | 198.673 | 800 | 27,896 | 21,245 | 223.693 |
| 410 | 13,699 | 10,290 | 199.521 | 810 | 28,284 | 21,549 | 224.174 |
| 420 | 14,043 | 10,551 | 200.350 | 820 | 28,672 | 21,855 | 224.651 |
| 430 | 14,388 | 10,813 | 201.160 | 830 | 29,062 | 22,162 | 225.123 |
| 440 | 14,734 | 11,075 | 201.955 | 840 | 29,454 | 22,470 | 225.592 |
| 450 | 15,080 | 11,339 | 202.734 | 850 | 29,846 | 22,779 | 226.057 |
| 460 | 15,428 | 11,603 | 203.497 | 860 | 30,240 | 23,090 | 226.517 |
| 470 | 15,777 | 11,869 | 204.247 | 870 | 30,635 | 23,402 | 226.973 |
| 480 | 16,126 | 12,135 | 204.982 | 880 | 31,032 | 23,715 | 227.426 |
| 490 | 16,477 | 12,403 | 205.705 | 890 | 31,429 | 24,029 | 227.875 |
| 500 | 16,828 | 12,671 | 206.413 | 900 | 31,828 | 24,345 | 228.321 |
| 510 | 17,181 | 12,940 | 207.112 | 910 | 32,228 | 24,662 | 228.763 |
| 520 | 17,534 | 13,211 | 207.799 | 920 | 32,629 | 24,980 | 229.202 |
| 530 | 17,889 | 13,482 | 208.475 | 930 | 33,032 | 25,300 | 229.637 |
| 540 | 18,245 | 13,755 | 209.139 | 940 | 33,436 | 25,621 | 230.070 |
| 550 | 18,601 | 14,028 | 209.795 | 950 | 33,841 | 25,943 | 230.499 |
| 560 | 18,959 | 14,303 | 210.440 | 960 | 34,247 | 26,265 | 230.924 |
| 570 | 19,318 | 14,579 | 211.075 | 970 | 34,653 | 26,588 | 231.347 |
| 580 | 19,678 | 14,856 | 211.702 | 980 | 35,061 | 26,913 | 231.767 |
| 590 | 20,039 | 15,134 | 212.320 | 990 | 35,472 | 27,240 | 232.184 |

Ideal-gas properties of water vapor, $H_2O$ (Concluded)

| $T$ K | $\bar{h}$ kJ/kmol | $\bar{u}$ kJ/kmol | $\bar{s}°$ kJ/kmol·K | $T$ K | $\bar{h}$ kJ/kmol | $\bar{u}$ kJ/kmol | $\bar{s}°$ kJ/kmol·K |
|---|---|---|---|---|---|---|---|
| 1000 | 35,882 | 27,568 | 232.597 | 1760 | 70,535 | 55,902 | 258.151 |
| 1020 | 36,709 | 28,228 | 233.415 | 1780 | 71,523 | 56,723 | 258.708 |
| 1040 | 37,542 | 28,895 | 234.223 | 1800 | 72,513 | 57,547 | 259.262 |
| 1060 | 38,380 | 29,567 | 235.020 | 1820 | 73,507 | 58,375 | 259.811 |
| 1080 | 39,223 | 30,243 | 235.806 | 1840 | 74,506 | 59,207 | 260.357 |
| 1100 | 40,071 | 30,925 | 236.584 | 1860 | 75,506 | 60,042 | 260.898 |
| 1120 | 40,923 | 31,611 | 237.352 | 1880 | 76,511 | 60,880 | 261.436 |
| 1140 | 41,780 | 32,301 | 238.110 | 1900 | 77,517 | 61,720 | 261.969 |
| 1160 | 42,642 | 32,997 | 238.859 | 1920 | 78,527 | 62,564 | 262.497 |
| 1180 | 43,509 | 33,698 | 239.600 | 1940 | 79,540 | 63,411 | 263.022 |
| 1200 | 44,380 | 34,403 | 240.333 | 1960 | 80,555 | 64,259 | 263.542 |
| 1220 | 45,256 | 35,112 | 241.057 | 1980 | 81,573 | 65,111 | 264.059 |
| 1240 | 46,137 | 35,827 | 241.773 | 2000 | 82,593 | 65,965 | 264.571 |
| 1260 | 47,022 | 36,546 | 242.482 | 2050 | 85,156 | 68,111 | 265.838 |
| 1280 | 47,912 | 37,270 | 243.183 | 2100 | 87,735 | 70,275 | 267.081 |
| 1300 | 48,807 | 38,000 | 243.877 | 2150 | 90,330 | 72,454 | 268.301 |
| 1320 | 49,707 | 38,732 | 244.564 | 2200 | 92,940 | 74,649 | 269.500 |
| 1340 | 50,612 | 39,470 | 245.243 | 2250 | 95,562 | 76,855 | 270.679 |
| 1360 | 51,521 | 40,213 | 245.915 | 2300 | 98,199 | 79,076 | 271.839 |
| 1380 | 52,434 | 40,960 | 246.582 | 2350 | 100,846 | 81,308 | 272.978 |
| 1400 | 53,351 | 41,711 | 247.241 | 2400 | 103,508 | 83,553 | 274.098 |
| 1420 | 54,273 | 42,466 | 247.895 | 2450 | 106,183 | 85,811 | 275.201 |
| 1440 | 55,198 | 43,226 | 248.543 | 2500 | 108,868 | 88,082 | 276.286 |
| 1460 | 56,128 | 43,989 | 249.185 | 2550 | 111,565 | 90,364 | 277.354 |
| 1480 | 57,062 | 44,756 | 249.820 | 2600 | 114,273 | 92,656 | 278.407 |
| 1500 | 57,999 | 45,528 | 250.450 | 2650 | 116,991 | 94,958 | 279.441 |
| 1520 | 58,942 | 46,304 | 251.074 | 2700 | 119,717 | 97,269 | 280.462 |
| 1540 | 59,888 | 47,084 | 251.693 | 2750 | 122,453 | 99,588 | 281.464 |
| 1560 | 60,838 | 47,868 | 252.305 | 2800 | 125,198 | 101,917 | 282.453 |
| 1580 | 61,792 | 48,655 | 252.912 | 2850 | 127,952 | 104,256 | 283.429 |
| 1600 | 62,748 | 49,445 | 253.513 | 2900 | 130,717 | 106,605 | 284.390 |
| 1620 | 63,709 | 50,240 | 254.111 | 2950 | 133,486 | 108,959 | 285.338 |
| 1640 | 64,675 | 51,039 | 254.703 | 3000 | 136,264 | 111,321 | 286.273 |
| 1660 | 65,643 | 51,841 | 255.290 | 3050 | 139,051 | 113,692 | 287.194 |
| 1680 | 66,614 | 52,646 | 255.873 | 3100 | 141,846 | 116,072 | 288.102 |
| 1700 | 67,589 | 53,455 | 256.450 | 3150 | 144,648 | 118,458 | 288.999 |
| 1720 | 68,567 | 54,267 | 257.022 | 3200 | 147,457 | 120,851 | 289.884 |
| 1740 | 69,550 | 55,083 | 257.589 | 3250 | 150,272 | 123,250 | 290.756 |

## TABLE A–24

Ideal-gas properties of monatomic oxygen, O

| $T$ K | $\overline{h}$ kJ/kmol | $\overline{u}$ kJ/kmol | $\overline{s}°$ kJ/kmol·K | $T$ K | $\overline{h}$ kJ/kmol | $\overline{u}$ kJ/kmol | $\overline{s}°$ kJ/kmol·K |
|---|---|---|---|---|---|---|---|
| 0 | 0 | 0 | 0 | 2400 | 50,894 | 30,940 | 204.932 |
| 298 | 6,852 | 4,373 | 160.944 | 2450 | 51,936 | 31,566 | 205.362 |
| 300 | 6,892 | 4,398 | 161.079 | 2500 | 52,979 | 32,193 | 205.783 |
| 500 | 11,197 | 7,040 | 172.088 | 2550 | 54,021 | 32,820 | 206.196 |
| 1000 | 21,713 | 13,398 | 186.678 | 2600 | 55,064 | 33,447 | 206.601 |
| 1500 | 32,150 | 19,679 | 195.143 | 2650 | 56,108 | 34,075 | 206.999 |
| 1600 | 34,234 | 20,931 | 196.488 | 2700 | 57,152 | 34,703 | 207.389 |
| 1700 | 36,317 | 22,183 | 197.751 | 2750 | 58,196 | 35,332 | 207.772 |
| 1800 | 38,400 | 23,434 | 198.941 | 2800 | 59,241 | 35,961 | 208.148 |
| 1900 | 40,482 | 24,685 | 200.067 | 2850 | 60,286 | 36,590 | 208.518 |
| 2000 | 42,564 | 25,935 | 201.135 | 2900 | 61,332 | 37,220 | 208.882 |
| 2050 | 43,605 | 26,560 | 201.649 | 2950 | 62,378 | 37,851 | 209.240 |
| 2100 | 44,646 | 27,186 | 202.151 | 3000 | 63,425 | 38,482 | 209.592 |
| 2150 | 45,687 | 27,811 | 202.641 | 3100 | 65,520 | 39,746 | 210.279 |
| 2200 | 46,728 | 28,436 | 203.119 | 3200 | 67,619 | 41,013 | 210.945 |
| 2250 | 47,769 | 29,062 | 203.588 | 3300 | 69,720 | 42,283 | 211.592 |
| 2300 | 48,811 | 29,688 | 204.045 | 3400 | 71,824 | 43,556 | 212.220 |
| 2350 | 49,852 | 30,314 | 204.493 | 3500 | 73,932 | 44,832 | 212.831 |

## TABLE A–25

Ideal-gas properties of hydroxyl, OH

| $T$ K | $\overline{h}$ kJ/kmol | $\overline{u}$ kJ/kmol | $\overline{s}°$ kJ/kmol·K | $T$ K | $\overline{h}$ kJ/kmol | $\overline{u}$ kJ/kmol | $\overline{s}°$ kJ/kmol·K |
|---|---|---|---|---|---|---|---|
| 0 | 0 | 0 | 0 | 2400 | 77,015 | 57,061 | 248.628 |
| 298 | 9,188 | 6,709 | 183.594 | 2450 | 78,801 | 58,431 | 249.364 |
| 300 | 9,244 | 6,749 | 183.779 | 2500 | 80,592 | 59,806 | 250.088 |
| 500 | 15,181 | 11,024 | 198.955 | 2550 | 82,388 | 61,186 | 250.799 |
| 1000 | 30,123 | 21,809 | 219.624 | 2600 | 84,189 | 62,572 | 251.499 |
| 1500 | 46,046 | 33,575 | 232.506 | 2650 | 85,995 | 63,962 | 252.187 |
| 1600 | 49,358 | 36,055 | 234.642 | 2700 | 87,806 | 65,358 | 252.864 |
| 1700 | 52,706 | 38,571 | 236.672 | 2750 | 89,622 | 66,757 | 253.530 |
| 1800 | 56,089 | 41,123 | 238.606 | 2800 | 91,442 | 68,162 | 254.186 |
| 1900 | 59,505 | 43,708 | 240.453 | 2850 | 93,266 | 69,570 | 254.832 |
| 2000 | 62,952 | 46,323 | 242.221 | 2900 | 95,095 | 70,983 | 255.468 |
| 2050 | 64,687 | 47,642 | 243.077 | 2950 | 96,927 | 72,400 | 256.094 |
| 2100 | 66,428 | 48,968 | 243.917 | 3000 | 98,763 | 73,820 | 256.712 |
| 2150 | 68,177 | 50,301 | 244.740 | 3100 | 102,447 | 76,673 | 257.919 |
| 2200 | 69,932 | 51,641 | 245.547 | 3200 | 106,145 | 79,539 | 259.093 |
| 2250 | 71,694 | 52,987 | 246.338 | 3300 | 109,855 | 82,418 | 260.235 |
| 2300 | 73,462 | 54,339 | 247.116 | 3400 | 113,578 | 85,309 | 261.347 |
| 2350 | 75,236 | 55,697 | 247.879 | 3500 | 117,312 | 88,212 | 262.429 |

## TABLE A–26

Enthalpy of formation, Gibbs function of formation, and absolute entropy at 25°C, 1 atm

| Substance | Formula | $\bar{h}_f^\circ$ kJ/kmol | $\bar{g}_f^\circ$ kJ/kmol | $\bar{s}^\circ$ kJ/kmol·K |
|---|---|---|---|---|
| Carbon | C($s$) | 0 | 0 | 5.74 |
| Hydrogen | H$_2$($g$) | 0 | 0 | 130.68 |
| Nitrogen | N$_2$($g$) | 0 | 0 | 191.61 |
| Oxygen | O$_2$($g$) | 0 | 0 | 205.04 |
| Carbon monoxide | CO($g$) | −110,530 | −137,150 | 197.65 |
| Carbon dioxide | CO$_2$($g$) | −393,520 | −394,360 | 213.80 |
| Water vapor | H$_2$O($g$) | −241,820 | −228,590 | 188.83 |
| Water | H$_2$O($l$) | −285,830 | −237,180 | 69.92 |
| Hydrogen peroxide | H$_2$O$_2$($g$) | −136,310 | −105,600 | 232.63 |
| Ammonia | NH$_3$($g$) | −46,190 | −16,590 | 192.33 |
| Methane | CH$_4$($g$) | −74,850 | −50,790 | 186.16 |
| Acetylene | C$_2$H$_2$($g$) | +226,730 | +209,170 | 200.85 |
| Ethylene | C$_2$H$_4$($g$) | +52,280 | +68,120 | 219.83 |
| Ethane | C$_2$H$_6$($g$) | −84,680 | −32,890 | 229.49 |
| Propylene | C$_3$H$_6$($g$) | +20,410 | +62,720 | 266.94 |
| Propane | C$_3$H$_8$($g$) | −103,850 | −23,490 | 269.91 |
| $n$-Butane | C$_4$H$_{10}$($g$) | −126,150 | −15,710 | 310.12 |
| $n$-Octane | C$_8$H$_{18}$($g$) | −208,450 | +16,530 | 466.73 |
| $n$-Octane | C$_8$H$_{18}$($l$) | −249,950 | +6,610 | 360.79 |
| $n$-Dodecane | C$_{12}$H$_{26}$($g$) | −291,010 | +50,150 | 622.83 |
| Benzene | C$_6$H$_6$($g$) | +82,930 | +129,660 | 269.20 |
| Methyl alcohol | CH$_3$OH($g$) | −200,670 | −162,000 | 239.70 |
| Methyl alcohol | CH$_3$OH($l$) | −238,660 | −166,360 | 126.80 |
| Ethyl alcohol | C$_2$H$_5$OH($g$) | −235,310 | −168,570 | 282.59 |
| Ethyl alcohol | C$_2$H$_5$OH($l$) | −277,690 | −174,890 | 160.70 |
| Oxygen | O($g$) | +249,190 | +231,770 | 161.06 |
| Hydrogen | H($g$) | +218,000 | +203,290 | 114.72 |
| Nitrogen | N($g$) | +472,650 | +455,510 | 153.30 |
| Hydroxyl | OH($g$) | +39,460 | +34,280 | 183.70 |

*Source of Data:* From JANAF, *Thermochemical Tables* (Midland, MI: Dow Chemical Co., 1971); *Selected Values of Chemical Thermodynamic Properties,* NBS Technical Note 270-3, 1968; and *API Research Project 44* (Carnegie Press, 1953).

## TABLE A-27

Properties of some common fuels and hydrocarbons

| Fuel (phase) | Formula | Molar mass, kg/kmol | Density,[1] kg/L | Enthalpy of vaporization,[2] kJ/kg | Specific heat,[1] $c_p$ kJ/kg·K | Higher heating value,[3] kJ/kg | Lower heating value,[3] kJ/kg |
|---|---|---|---|---|---|---|---|
| Carbon (s) | C | 12.011 | 2 | — | 0.708 | 32,800 | 32,800 |
| Hydrogen (g) | $H_2$ | 2.016 | — | — | 14.4 | 141,800 | 120,000 |
| Carbon monoxide (g) | CO | 28.013 | — | — | 1.05 | 10,100 | 10,100 |
| Methane (g) | $CH_4$ | 16.043 | — | 509 | 2.20 | 55,530 | 50,050 |
| Methanol (l) | $CH_4O$ | 32.042 | 0.790 | 1168 | 2.53 | 22,660 | 19,920 |
| Acetylene (g) | $C_2H_2$ | 26.038 | — | — | 1.69 | 49,970 | 48,280 |
| Ethane (g) | $C_2H_6$ | 30.070 | — | 172 | 1.75 | 51,900 | 47,520 |
| Ethanol (l) | $C_2H_6O$ | 46.069 | 0.790 | 919 | 2.44 | 29,670 | 26,810 |
| Propane (l) | $C_3H_8$ | 44.097 | 0.500 | 335 | 2.77 | 50,330 | 46,340 |
| Butane (l) | $C_4H_{10}$ | 58.123 | 0.579 | 362 | 2.42 | 49,150 | 45,370 |
| 1-Pentene (l) | $C_5H_{10}$ | 70.134 | 0.641 | 363 | 2.20 | 47,760 | 44,630 |
| Isopentane (l) | $C_5H_{12}$ | 72.150 | 0.626 | — | 2.32 | 48,570 | 44,910 |
| Benzene (l) | $C_6H_6$ | 78.114 | 0.877 | 433 | 1.72 | 41,800 | 40,100 |
| Hexene (l) | $C_6H_{12}$ | 84.161 | 0.673 | 392 | 1.84 | 47,500 | 44,400 |
| Hexane (l) | $C_6H_{14}$ | 86.177 | 0.660 | 366 | 2.27 | 48,310 | 44,740 |
| Toluene (l) | $C_7H_8$ | 92.141 | 0.867 | 412 | 1.71 | 42,400 | 40,500 |
| Heptane (l) | $C_7H_{16}$ | 100.204 | 0.684 | 365 | 2.24 | 48,100 | 44,600 |
| Octane (l) | $C_8H_{18}$ | 114.231 | 0.703 | 363 | 2.23 | 47,890 | 44,430 |
| Decane (l) | $C_{10}H_{22}$ | 142.285 | 0.730 | 361 | 2.21 | 47,640 | 44,240 |
| Gasoline (l) | $C_nH_{1.87n}$ | 100–110 | 0.72–0.78 | 350 | 2.4 | 47,300 | 44,000 |
| Light diesel (l) | $C_nH_{1.8n}$ | 170 | 0.78–0.84 | 270 | 2.2 | 46,100 | 43,200 |
| Heavy diesel (l) | $C_nH_{1.7n}$ | 200 | 0.82–0.88 | 230 | 1.9 | 45,500 | 42,800 |
| Natural gas (g) | $C_nH_{3.8n}N_{0.1n}$ | 18 | — | — | 2 | 50,000 | 45,000 |

[1] At 1 atm and 20°C.
[2] At 25°C for liquid fuels, and 1 atm and normal boiling temperature for gaseous fuels.
[3] At 25°C. Multiply by molar mass to obtain heating values in kJ/kmol.

## TABLE A–28

Natural logarithms of the equilibrium constant $K_p$

The equilibrium constant $K_p$ for the reaction $\nu_A A + \nu_B B \rightleftharpoons \nu_C C + \nu_D D$ is defined as $K_p \equiv \dfrac{P_C^{\nu_C} P_D^{\nu_D}}{P_A^{\nu_A} P_B^{\nu_B}}$

| Temp., K | $H_2 \rightleftharpoons 2H$ | $O_2 \rightleftharpoons 2O$ | $N_2 \rightleftharpoons 2N$ | $H_2O \rightleftharpoons H_2 + \frac{1}{2}O_2$ | $H_2O \rightleftharpoons \frac{1}{2}H_2 + OH$ | $CO_2 \rightleftharpoons CO + \frac{1}{2}O_2$ | $\frac{1}{2}N_2 + \frac{1}{2}O_2 \rightleftharpoons NO$ |
|---|---|---|---|---|---|---|---|
| 298 | −164.005 | −186.975 | −367.480 | −92.208 | −106.208 | −103.762 | −35.052 |
| 500 | −92.827 | −105.630 | −213.372 | −52.691 | −60.281 | −57.616 | −20.295 |
| 1000 | −39.803 | −45.150 | −99.127 | −23.163 | −26.034 | −23.529 | −9.388 |
| 1200 | −30.874 | −35.005 | −80.011 | −18.182 | −20.283 | −17.871 | −7.569 |
| 1400 | −24.463 | −27.742 | −66.329 | −14.609 | −16.099 | −13.842 | −6.270 |
| 1600 | −19.637 | −22.285 | −56.055 | −11.921 | −13.066 | −10.830 | −5.294 |
| 1800 | −15.866 | −18.030 | −48.051 | −9.826 | −10.657 | −8.497 | −4.536 |
| 2000 | −12.840 | −14.622 | −41.645 | −8.145 | −8.728 | −6.635 | −3.931 |
| 2200 | −10.353 | −11.827 | −36.391 | −6.768 | −7.148 | −5.120 | −3.433 |
| 2400 | −8.276 | −9.497 | −32.011 | −5.619 | −5.832 | −3.860 | −3.019 |
| 2600 | −6.517 | −7.521 | −28.304 | −4.648 | −4.719 | −2.801 | −2.671 |
| 2800 | −5.002 | −5.826 | −25.117 | −3.812 | −3.763 | −1.894 | −2.372 |
| 3000 | −3.685 | −4.357 | −22.359 | −3.086 | −2.937 | −1.111 | −2.114 |
| 3200 | −2.534 | −3.072 | −19.937 | −2.451 | −2.212 | −0.429 | −1.888 |
| 3400 | −1.516 | −1.935 | −17.800 | −1.891 | −1.576 | 0.169 | −1.690 |
| 3600 | −0.609 | −0.926 | −15.898 | −1.392 | −1.088 | 0.701 | −1.513 |
| 3800 | 0.202 | −0.019 | −14.199 | −0.945 | −0.501 | 1.176 | −1.356 |
| 4000 | 0.934 | 0.796 | −12.660 | −0.542 | −0.044 | 1.599 | −1.216 |
| 4500 | 2.486 | 2.513 | −9.414 | 0.312 | 0.920 | 2.490 | −0.921 |
| 5000 | 3.725 | 3.895 | −6.807 | 0.996 | 1.689 | 3.197 | −0.686 |
| 5500 | 4.743 | 5.023 | −4.666 | 1.560 | 2.318 | 3.771 | −0.497 |
| 6000 | 5.590 | 5.963 | −2.865 | 2.032 | 2.843 | 4.245 | −0.341 |

*Source of Data:* Gordon J. Van Wylen and Richard E. Sonntag, *Fundamentals of Classical Thermodynamics,* English/SI Version, 3rd ed. (New York: John Wiley & Sons, 1986), p. 723, table A.14. Based on thermodynamic data given in JANAF, *Thermochemical Tables* (Midland, MI: Thermal Research Laboratory, The Dow Chemical Company, 1971).

**FIGURE A–29**

Generalized enthalpy departure chart.

*Source of Data: Redrawn from Gordon van Wylen and Richard Sontag,* Fundamentals of Classical Thermodynamics, *(SI version), 2d ed., Wiley, New York, 1976.*

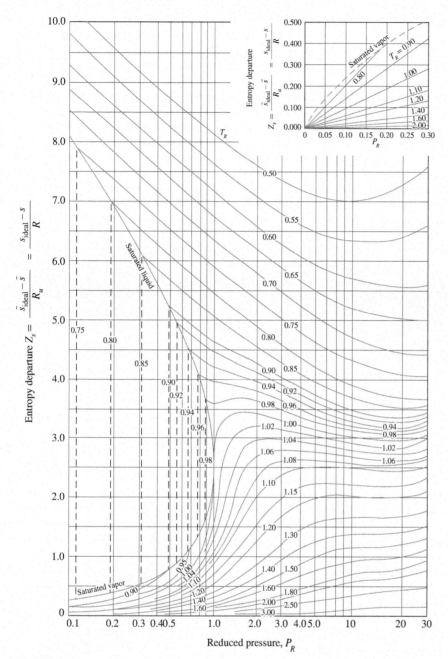

**FIGURE A–30**

Generalized entropy departure chart.

*Source of Data: Redrawn from Gordon van Wylen and Richard Sontag,* Fundamentals of Classical Thermodynamics, *(SI version), 2d ed., Wiley, New York, 1976.*

ASHRAE Psychrometric Chart No. 1
Normal Temperature
Barometric Pressure: 101.325 kPa

©1992 American Society of Heating,
Refrigerating and Air-Conditioning Engineers, Inc.

Prepared by Center for Applied Thermodynamic Studies, University of Idaho.

Sea Level

$$\text{Sensible heat} = \frac{DH_S}{Total\ heat} = \frac{DH_S}{DH_T}$$

$$\frac{\text{Enthalpy}}{\text{Humidity ratio}} = \frac{Dh}{D\omega}$$

Humidity ratio (ω) grams moisture per kilogram dry air

Enthalpy (h) kilojoules per kilogram dry air

Saturation temperature °C

Dry bulb temperature °C

25 wet bulb temperature °C

0.92 volume cubic meter per kilogram dry air

**FIGURE A–31**

Psychrometric chart at 1 atm total pressure.

*Reprinted from American Society of Heating, Refrigerating and Air-Conditioning Engineers, Inc., Atlanta, GA.*

## TABLE A–32

One-dimensional isentropic compressible-flow functions for an ideal gas with $k = 1.4$

| Ma | Ma* | $A/A^*$ | $P/P_0$ | $\rho/\rho_0$ | $T/T_0$ |
|-----|--------|---------|---------|---------|---------|
| 0 | 0 | $\infty$ | 1.0000 | 1.0000 | 1.0000 |
| 0.1 | 0.1094 | 5.8218 | 0.9930 | 0.9950 | 0.9980 |
| 0.2 | 0.2182 | 2.9635 | 0.9725 | 0.9803 | 0.9921 |
| 0.3 | 0.3257 | 2.0351 | 0.9395 | 0.9564 | 0.9823 |
| 0.4 | 0.4313 | 1.5901 | 0.8956 | 0.9243 | 0.9690 |
| 0.5 | 0.5345 | 1.3398 | 0.8430 | 0.8852 | 0.9524 |
| 0.6 | 0.6348 | 1.1882 | 0.7840 | 0.8405 | 0.9328 |
| 0.7 | 0.7318 | 1.0944 | 0.7209 | 0.7916 | 0.9107 |
| 0.8 | 0.8251 | 1.0382 | 0.6560 | 0.7400 | 0.8865 |
| 0.9 | 0.9146 | 1.0089 | 0.5913 | 0.6870 | 0.8606 |
| 1.0 | 1.0000 | 1.0000 | 0.5283 | 0.6339 | 0.8333 |
| 1.2 | 1.1583 | 1.0304 | 0.4124 | 0.5311 | 0.7764 |
| 1.4 | 1.2999 | 1.1149 | 0.3142 | 0.4374 | 0.7184 |
| 1.6 | 1.4254 | 1.2502 | 0.2353 | 0.3557 | 0.6614 |
| 1.8 | 1.5360 | 1.4390 | 0.1740 | 0.2868 | 0.6068 |
| 2.0 | 1.6330 | 1.6875 | 0.1278 | 0.2300 | 0.5556 |
| 2.2 | 1.7179 | 2.0050 | 0.0935 | 0.1841 | 0.5081 |
| 2.4 | 1.7922 | 2.4031 | 0.0684 | 0.1472 | 0.4647 |
| 2.6 | 1.8571 | 2.8960 | 0.0501 | 0.1179 | 0.4252 |
| 2.8 | 1.9140 | 3.5001 | 0.0368 | 0.0946 | 0.3894 |
| 3.0 | 1.9640 | 4.2346 | 0.0272 | 0.0760 | 0.3571 |
| 5.0 | 2.2361 | 25.000 | 0.0019 | 0.0113 | 0.1667 |
| $\infty$ | 2.2495 | $\infty$ | 0 | 0 | 0 |

$$\text{Ma}^* = \text{Ma}\sqrt{\frac{k+1}{2+(k-1)\text{Ma}^2}}$$

$$\frac{A}{A^*} = \frac{1}{\text{Ma}}\left[\left(\frac{2}{k+1}\right)\left(1+\frac{k-1}{2}\text{Ma}^2\right)\right]^{0.5(k+1)/(k-1)}$$

$$\frac{P}{P_0} = \left(1+\frac{k-1}{2}\text{Ma}^2\right)^{-k/(k-1)}$$

$$\frac{\rho}{\rho_0} = \left(1+\frac{k-1}{2}\text{Ma}^2\right)^{-1/(k-1)}$$

$$\frac{T}{T_0} = \left(1+\frac{k-1}{2}\text{Ma}^2\right)^{-1}$$

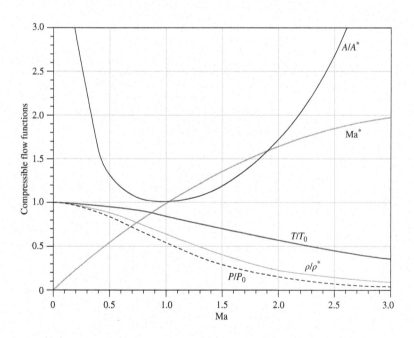

## TABLE A–33

One-dimensional normal-shock functions for an ideal gas with $k = 1.4$

| $Ma_1$ | $Ma_2$ | $P_2/P_1$ | $\rho_2/\rho_1$ | $T_2/T_1$ | $P_{02}/P_{01}$ | $P_{02}/P_1$ |
|---|---|---|---|---|---|---|
| 1.0 | 1.0000 | 1.0000 | 1.0000 | 1.0000 | 1.0000 | 1.8929 |
| 1.1 | 0.9118 | 1.2450 | 1.1691 | 1.0649 | 0.9989 | 2.1328 |
| 1.2 | 0.8422 | 1.5133 | 1.3416 | 1.1280 | 0.9928 | 2.4075 |
| 1.3 | 0.7860 | 1.8050 | 1.5157 | 1.1909 | 0.9794 | 2.7136 |
| 1.4 | 0.7397 | 2.1200 | 1.6897 | 1.2547 | 0.9582 | 3.0492 |
| 1.5 | 0.7011 | 2.4583 | 1.8621 | 1.3202 | 0.9298 | 3.4133 |
| 1.6 | 0.6684 | 2.8200 | 2.0317 | 1.3880 | 0.8952 | 3.8050 |
| 1.7 | 0.6405 | 3.2050 | 2.1977 | 1.4583 | 0.8557 | 4.2238 |
| 1.8 | 0.6165 | 3.6133 | 2.3592 | 1.5316 | 0.8127 | 4.6695 |
| 1.9 | 0.5956 | 4.0450 | 2.5157 | 1.6079 | 0.7674 | 5.1418 |
| 2.0 | 0.5774 | 4.5000 | 2.6667 | 1.6875 | 0.7209 | 5.6404 |
| 2.1 | 0.5613 | 4.9783 | 2.8119 | 1.7705 | 0.6742 | 6.1654 |
| 2.2 | 0.5471 | 5.4800 | 2.9512 | 1.8569 | 0.6281 | 6.7165 |
| 2.3 | 0.5344 | 6.0050 | 3.0845 | 1.9468 | 0.5833 | 7.2937 |
| 2.4 | 0.5231 | 6.5533 | 3.2119 | 2.0403 | 0.5401 | 7.8969 |
| 2.5 | 0.5130 | 7.1250 | 3.3333 | 2.1375 | 0.4990 | 8.5261 |
| 2.6 | 0.5039 | 7.7200 | 3.4490 | 2.2383 | 0.4601 | 9.1813 |
| 2.7 | 0.4956 | 8.3383 | 3.5590 | 2.3429 | 0.4236 | 9.8624 |
| 2.8 | 0.4882 | 8.9800 | 3.6636 | 2.4512 | 0.3895 | 10.5694 |
| 2.9 | 0.4814 | 9.6450 | 3.7629 | 2.5632 | 0.3577 | 11.3022 |
| 3.0 | 0.4752 | 10.3333 | 3.8571 | 2.6790 | 0.3283 | 12.0610 |
| 4.0 | 0.4350 | 18.5000 | 4.5714 | 4.0469 | 0.1388 | 21.0681 |
| 5.0 | 0.4152 | 29.000 | 5.0000 | 5.8000 | 0.0617 | 32.6335 |
| $\infty$ | 0.3780 | $\infty$ | 6.0000 | $\infty$ | 0 | $\infty$ |

$$T_{01} = T_{02}$$

$$Ma_2 = \sqrt{\frac{(k-1)Ma_1^2 + 2}{2kMa_1^2 - k + 1}}$$

$$\frac{P_2}{P_1} = \frac{1 + kMa_1^2}{1 + kMa_2^2} = \frac{2kMa_1^2 - k + 1}{k+1}$$

$$\frac{\rho_2}{\rho_1} = \frac{P_2/P_1}{T_2/T_1} = \frac{(k+1)Ma_1^2}{2 + (k-1)Ma_1^2} = \frac{V_1}{V_2}$$

$$\frac{T_2}{T_1} = \frac{2 + Ma_1^2(k-1)}{2 + Ma_1^2(k-1)}$$

$$\frac{P_{02}}{P_{01}} = \frac{Ma_1}{Ma_2}\left[\frac{1 + Ma_2^2(k-1)/2}{1 + Ma_1^2(k-1)/2}\right]^{(k+1)[2(k-1)]}$$

$$\frac{P_{02}}{P_{01}} = \frac{(1 + kMa_1^2)\left[1 + Ma_2^2(k-1)/2\right]^{k/(k-1)}}{1 + kMa_2^2}$$

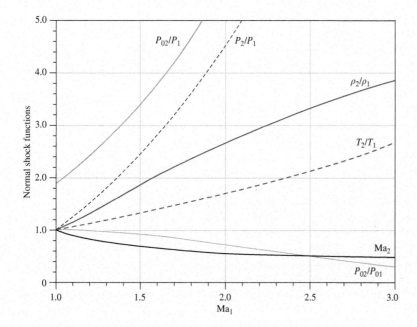

Rayleigh flow functions for an ideal gas with $k = 1.4$

| Ma | $T_0/T_0^*$ | $P_0/P_0^*$ | $T/T^*$ | $P/P^*$ | $V/V^*$ |
|---|---|---|---|---|---|
| 0.0 | 0.0000 | 1.2679 | 0.0000 | 2.4000 | 0.0000 |
| 0.1 | 0.0468 | 1.2591 | 0.0560 | 2.3669 | 0.0237 |
| 0.2 | 0.1736 | 1.2346 | 0.2066 | 2.2727 | 0.0909 |
| 0.3 | 0.3469 | 1.1985 | 0.4089 | 2.1314 | 0.1918 |
| 0.4 | 0.5290 | 1.1566 | 0.6151 | 1.9608 | 0.3137 |
| 0.5 | 0.6914 | 1.1141 | 0.7901 | 1.7778 | 0.4444 |
| 0.6 | 0.8189 | 1.0753 | 0.9167 | 1.5957 | 0.5745 |
| 0.7 | 0.9085 | 1.0431 | 0.9929 | 1.4235 | 0.6975 |
| 0.8 | 0.9639 | 1.0193 | 1.0255 | 1.2658 | 0.8101 |
| 0.9 | 0.9921 | 1.0049 | 1.0245 | 1.1246 | 0.9110 |
| 1.0 | 1.0000 | 1.0000 | 1.0000 | 1.0000 | 1.0000 |
| 1.2 | 0.9787 | 1.0194 | 0.9118 | 0.7958 | 1.1459 |
| 1.4 | 0.9343 | 1.0777 | 0.8054 | 0.6410 | 1.2564 |
| 1.6 | 0.8842 | 1.1756 | 0.7017 | 0.5236 | 1.3403 |
| 1.8 | 0.8363 | 1.3159 | 0.6089 | 0.4335 | 1.4046 |
| 2.0 | 0.7934 | 1.5031 | 0.5289 | 0.3636 | 1.4545 |
| 2.2 | 0.7561 | 1.7434 | 0.4611 | 0.3086 | 1.4938 |
| 2.4 | 0.7242 | 2.0451 | 0.4038 | 0.2648 | 1.5252 |
| 2.6 | 0.6970 | 2.4177 | 0.3556 | 0.2294 | 1.5505 |
| 2.8 | 0.6738 | 2.8731 | 0.3149 | 0.2004 | 1.5711 |
| 3.0 | 0.6540 | 3.4245 | 0.2803 | 0.1765 | 1.5882 |

$$\frac{T_0}{T_0^*} = \frac{(k+1)\mathrm{Ma}^2\left[2 + (k-1)\mathrm{Ma}^2\right]}{\left(1 + k\mathrm{Ma}^2\right)^2}$$

$$\frac{P_0}{P_0^*} = \frac{k+1}{1 + k\mathrm{Ma}^2}\left(\frac{2 + (k-1)\mathrm{Ma}^2}{k+1}\right)^{k/(k-1)}$$

$$\frac{T}{T^*} = \left(\frac{\mathrm{Ma}(1+k)}{1 + k\mathrm{Ma}^2}\right)^2$$

$$\frac{P}{P^*} = \frac{1+k}{1 + k\mathrm{Ma}^2}$$

$$\frac{V}{V^*} = \frac{\rho^*}{\rho} = \frac{(1+k)\mathrm{Ma}^2}{1 + k\mathrm{Ma}^2}$$

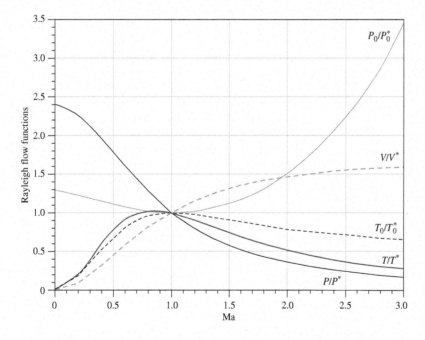

# 2

# PROPERTY TABLES AND CHARTS (ENGLISH UNITS)

Molar mass, gas constant, and critical-point properties

| Substance | Formula | Molar mass, $M$ lbm/lbmol | Gas constant, $R^*$ | | Critical-point properties | | |
|---|---|---|---|---|---|---|---|
| | | | Btu/ lbm·R | psia·ft³/ lbm·R | Temperature, R | Pressure, psia | Volume, ft³/lbmol |
| Air | — | 28.97 | 0.06855 | 0.3704 | 238.5 | 547 | 1.41 |
| Ammonia | $NH_3$ | 17.03 | 0.1166 | 0.6301 | 729.8 | 1636 | 1.16 |
| Argon | Ar | 39.948 | 0.04971 | 0.2686 | 272 | 705 | 1.20 |
| Benzene | $C_6H_6$ | 78.115 | 0.02542 | 0.1374 | 1012 | 714 | 4.17 |
| Bromine | $Br_2$ | 159.808 | 0.01243 | 0.06714 | 1052 | 1500 | 2.17 |
| $n$-Butane | $C_4H_{10}$ | 58.124 | 0.03417 | 0.1846 | 765.2 | 551 | 4.08 |
| Carbon dioxide | $CO_2$ | 44.01 | 0.04513 | 0.2438 | 547.5 | 1071 | 1.51 |
| Carbon monoxide | CO | 28.011 | 0.07090 | 0.3831 | 240 | 507 | 1.49 |
| Carbon tetrachloride | $CCl_4$ | 153.82 | 0.01291 | 0.06976 | 1001.5 | 661 | 4.42 |
| Chlorine | $Cl_2$ | 70.906 | 0.02801 | 0.1517 | 751 | 1120 | 1.99 |
| Chloroform | $CHCl_3$ | 119.38 | 0.01664 | 0.08988 | 965.8 | 794 | 3.85 |
| Dichlorodifluoromethane (R-12) | $CCl_2F_2$ | 120.91 | 0.01643 | 0.08874 | 692.4 | 582 | 3.49 |
| Dichlorofluoromethane (R-21) | $CHCl_2F$ | 102.92 | 0.01930 | 0.1043 | 813.0 | 749 | 3.16 |
| Ethane | $C_2H_6$ | 30.020 | 0.06616 | 0.3574 | 549.8 | 708 | 2.37 |
| Ethyl alcohol | $C_2H_5OH$ | 46.07 | 0.04311 | 0.2329 | 929.0 | 926 | 2.68 |
| Ethylene | $C_2H_4$ | 28.054 | 0.07079 | 0.3825 | 508.3 | 742 | 1.99 |
| Helium | He | 4.003 | 0.4961 | 2.6809 | 9.5 | 33.2 | 0.926 |
| $n$-Hexane | $C_6H_{14}$ | 86.178 | 0.02305 | 0.1245 | 914.2 | 439 | 5.89 |
| Hydrogen (normal) | $H_2$ | 2.016 | 0.9851 | 5.3224 | 59.9 | 188.1 | 1.04 |
| Krypton | Kr | 83.80 | 0.02370 | 0.1280 | 376.9 | 798 | 1.48 |
| Methane | $CH_4$ | 16.043 | 0.1238 | 0.6688 | 343.9 | 673 | 1.59 |
| Methyl alcohol | $CH_3OH$ | 32.042 | 0.06198 | 0.3349 | 923.7 | 1154 | 1.89 |
| Methyl chloride | $CH_3Cl$ | 50.488 | 0.03934 | 0.2125 | 749.3 | 968 | 2.29 |
| Neon | Ne | 20.183 | 0.09840 | 0.5316 | 80.1 | 395 | 0.668 |
| Nitrogen | $N_2$ | 28.013 | 0.07090 | 0.3830 | 227.1 | 492 | 1.44 |
| Nitrous oxide | $N_2O$ | 44.013 | 0.04512 | 0.2438 | 557.4 | 1054 | 1.54 |
| Oxygen | $O_2$ | 31.999 | 0.06206 | 0.3353 | 278.6 | 736 | 1.25 |
| Propane | $C_3H_8$ | 44.097 | 0.04504 | 0.2433 | 665.9 | 617 | 3.20 |
| Propylene | $C_3H_6$ | 42.081 | 0.04719 | 0.2550 | 656.9 | 670 | 2.90 |
| Sulfur dioxide | $SO_2$ | 64.063 | 0.03100 | 1.1675 | 775.2 | 1143 | 1.95 |
| Tetrafluoroethane (R-134a) | $CF_3CH_2F$ | 102.03 | 0.01946 | 0.1052 | 673.6 | 588.7 | 3.19 |
| Trichlorofluoromethane (R-11) | $CCl_3F$ | 137.37 | 0.01446 | 0.07811 | 848.1 | 635 | 3.97 |
| Water | $H_2O$ | 18.015 | 0.1102 | 0.5956 | 1164.8 | 3200 | 0.90 |
| Xenon | Xe | 131.30 | 0.01513 | 0.08172 | 521.55 | 852 | 1.90 |

*Calculated from $R = R_u/M$, where $R_u = 1.98588$ Btu/lbmol·R = 10.7316 psia·ft³/lbmol·R and $M$ is the molar mass.

*Source of Data:* K. A. Kobe and R. E. Lynn, Jr., *Chemical Review* 52 (1953), pp. 117–236, and ASHRAE, *Handbook of Fundamentals* (Atlanta, GA: American Society of Heating, Refrigerating, and Air-Conditioning Engineers, Inc., 1993), pp. 16.4 and 36.1.

Ideal-gas specific heats of various common gases

(a) At 80°F

| Gas | Formula | Gas constant, $R$ Btu/lbm·R | $c_p$ Btu/lbm·R | $c_v$ Btu/lbm·R | $k$ |
|---|---|---|---|---|---|
| Air | — | 0.06855 | 0.240 | 0.171 | 1.400 |
| Argon | Ar | 0.04971 | 0.1253 | 0.0756 | 1.667 |
| Butane | $C_4H_{10}$ | 0.03424 | 0.415 | 0.381 | 1.09 |
| Carbon dioxide | $CO_2$ | 0.04513 | 0.203 | 0.158 | 1.285 |
| Carbon monoxide | CO | 0.07090 | 0.249 | 0.178 | 1.399 |
| Ethane | $C_2H_6$ | 0.06616 | 0.427 | 0.361 | 1.183 |
| Ethylene | $C_2H_4$ | 0.07079 | 0.411 | 0.340 | 1.208 |
| Helium | He | 0.4961 | 1.25 | 0.753 | 1.667 |
| Hydrogen | $H_2$ | 0.9851 | 3.43 | 2.44 | 1.404 |
| Methane | $CH_4$ | 0.1238 | 0.532 | 0.403 | 1.32 |
| Neon | Ne | 0.09840 | 0.246 | 0.1477 | 1.667 |
| Nitrogen | $N_2$ | 0.07090 | 0.248 | 0.177 | 1.400 |
| Octane | $C_8H_{18}$ | 0.01742 | 0.409 | 0.392 | 1.044 |
| Oxygen | $O_2$ | 0.06206 | 0.219 | 0.157 | 1.395 |
| Propane | $C_3H_8$ | 0.04504 | 0.407 | 0.362 | 1.124 |
| Steam | $H_2O$ | 0.1102 | 0.445 | 0.335 | 1.329 |

*Source of Data:* Gordon J. Van Wylen and Richard E. Sonntag, *Fundamentals of Classical Thermodynamics,* English/SI Version, 3rd ed. (New York: John Wiley & Sons, 1986), p. 687, Table A–8E.

## TABLE A–2E

Ideal-gas specific heats of various common gases (*Continued*)
(*b*) At various temperatures

| Temp., °F | $c_p$ Btu/lbm·R | $c_v$ Btu/lbm·R | $k$ | $c_p$ Btu/lbm·R | $c_v$ Btu/lbm·R | $k$ | $c_p$ Btu/lbm·R | $c_v$ Btu/lbm·R | $k$ |
|---|---|---|---|---|---|---|---|---|---|
| | *Air* | | | *Carbon dioxide, $CO_2$* | | | *Carbon monoxide, CO* | | |
| 40 | 0.240 | 0.171 | 1.401 | 0.195 | 0.150 | 1.300 | 0.248 | 0.177 | 1.400 |
| 100 | 0.240 | 0.172 | 1.400 | 0.205 | 0.160 | 1.283 | 0.249 | 0.178 | 1.399 |
| 200 | 0.241 | 0.173 | 1.397 | 0.217 | 0.172 | 1.262 | 0.249 | 0.179 | 1.397 |
| 300 | 0.243 | 0.174 | 1.394 | 0.229 | 0.184 | 1.246 | 0.251 | 0.180 | 1.394 |
| 400 | 0.245 | 0.176 | 1.389 | 0.239 | 0.193 | 1.233 | 0.253 | 0.182 | 1.389 |
| 500 | 0.248 | 0.179 | 1.383 | 0.247 | 0.202 | 1.223 | 0.256 | 0.185 | 1.384 |
| 600 | 0.250 | 0.182 | 1.377 | 0.255 | 0.210 | 1.215 | 0.259 | 0.188 | 1.377 |
| 700 | 0.254 | 0.185 | 1.371 | 0.262 | 0.217 | 1.208 | 0.262 | 0.191 | 1.371 |
| 800 | 0.257 | 0.188 | 1.365 | 0.269 | 0.224 | 1.202 | 0.266 | 0.195 | 1.364 |
| 900 | 0.259 | 0.191 | 1.358 | 0.275 | 0.230 | 1.197 | 0.269 | 0.198 | 1.357 |
| 1000 | 0.263 | 0.195 | 1.353 | 0.280 | 0.235 | 1.192 | 0.273 | 0.202 | 1.351 |
| 1500 | 0.276 | 0.208 | 1.330 | 0.298 | 0.253 | 1.178 | 0.287 | 0.216 | 1.328 |
| 2000 | 0.286 | 0.217 | 1.312 | 0.312 | 0.267 | 1.169 | 0.297 | 0.226 | 1.314 |
| | *Hydrogen, $H_2$* | | | *Nitrogen, $N_2$* | | | *Oxygen, $O_2$* | | |
| 40 | 3.397 | 2.412 | 1.409 | 0.248 | 0.177 | 1.400 | 0.219 | 0.156 | 1.397 |
| 100 | 3.426 | 2.441 | 1.404 | 0.248 | 0.178 | 1.399 | 0.220 | 0.158 | 1.394 |
| 200 | 3.451 | 2.466 | 1.399 | 0.249 | 0.178 | 1.398 | 0.223 | 0.161 | 1.387 |
| 300 | 3.461 | 2.476 | 1.398 | 0.250 | 0.179 | 1.396 | 0.226 | 0.164 | 1.378 |
| 400 | 3.466 | 2.480 | 1.397 | 0.251 | 0.180 | 1.393 | 0.230 | 0.168 | 1.368 |
| 500 | 3.469 | 2.484 | 1.397 | 0.254 | 0.183 | 1.388 | 0.235 | 0.173 | 1.360 |
| 600 | 3.473 | 2.488 | 1.396 | 0.256 | 0.185 | 1.383 | 0.239 | 0.177 | 1.352 |
| 700 | 3.477 | 2.492 | 1.395 | 0.260 | 0.189 | 1.377 | 0.242 | 0.181 | 1.344 |
| 800 | 3.494 | 2.509 | 1.393 | 0.262 | 0.191 | 1.371 | 0.246 | 0.184 | 1.337 |
| 900 | 3.502 | 2.519 | 1.392 | 0.265 | 0.194 | 1.364 | 0.249 | 0.187 | 1.331 |
| 1000 | 3.513 | 2.528 | 1.390 | 0.269 | 0.198 | 1.359 | 0.252 | 0.190 | 1.326 |
| 1500 | 3.618 | 2.633 | 1.374 | 0.283 | 0.212 | 1.334 | 0.263 | 0.201 | 1.309 |
| 2000 | 3.758 | 2.773 | 1.355 | 0.293 | 0.222 | 1.319 | 0.270 | 0.208 | 1.298 |

*Note:* The unit Btu/lbm·R is equivalent to Btu/lbm°F.

*Source of Data:* Kenneth Wark, *Thermodynamics,* 4th ed. (New York: McGraw-Hill, 1983), p. 830, Table A–4. Originally published in *Tables of Properties of Gases,* NBS Circular 564, 1955.

Ideal-gas specific heats of various common gases (*Concluded*)
(*c*) As a function of temperature

$$\overline{c}_p = a + bT + cT^2 + dT^3$$

($T$ in R, $c_p$ in Btu/lbmol·R)

| Substance | Formula | $a$ | $b$ | $c$ | $d$ | Temperature range, R | % error Max. | % error Avg. |
|---|---|---|---|---|---|---|---|---|
| Nitrogen | $N_2$ | 6.903 | $-0.02085 \times 10^{-2}$ | $0.05957 \times 10^{-5}$ | $-0.1176 \times 10^{-9}$ | 491–3240 | 0.59 | 0.34 |
| Oxygen | $O_2$ | 6.085 | $0.2017 \times 10^{-2}$ | $-0.05275 \times 10^{-5}$ | $0.05372 \times 10^{-9}$ | 491–3240 | 1.19 | 0.28 |
| Air | — | 6.713 | $0.02609 \times 10^{-2}$ | $0.03540 \times 10^{-5}$ | $-0.08052 \times 10^{-9}$ | 491–3240 | 0.72 | 0.33 |
| Hydrogen | $H_2$ | 6.952 | $-0.02542 \times 10^{-2}$ | $0.02952 \times 10^{-5}$ | $-0.03565 \times 10^{-9}$ | 491–3240 | 1.02 | 0.26 |
| Carbon monoxide | CO | 6.726 | $0.02222 \times 10^{-2}$ | $0.03960 \times 10^{-5}$ | $-0.09100 \times 10^{-9}$ | 491–3240 | 0.89 | 0.37 |
| Carbon dioxide | $CO_2$ | 5.316 | $0.79361 \times 10^{-2}$ | $-0.2581 \times 10^{-5}$ | $0.3059 \times 10^{-9}$ | 491–3240 | 0.67 | 0.22 |
| Water vapor | $H_2O$ | 7.700 | $0.02552 \times 10^{-2}$ | $0.07781 \times 10^{-5}$ | $-0.1472 \times 10^{-9}$ | 491–3240 | 0.53 | 0.24 |
| Nitric oxide | NO | 7.008 | $-0.01247 \times 10^{-2}$ | $0.07185 \times 10^{-5}$ | $-0.1715 \times 10^{-9}$ | 491–2700 | 0.97 | 0.36 |
| Nitrous oxide | $N_2O$ | 5.758 | $0.7780 \times 10^{-2}$ | $-0.2596 \times 10^{-5}$ | $0.4331 \times 10^{-9}$ | 491–2700 | 0.59 | 0.26 |
| Nitrogen dioxide | $NO_2$ | 5.48 | $0.7583 \times 10^{-2}$ | $-0.260 \times 10^{-5}$ | $0.322 \times 10^{-9}$ | 491–2700 | 0.46 | 0.18 |
| Ammonia | $NH_3$ | 6.5846 | $0.34028 \times 10^{-2}$ | $0.073034 \times 10^{-5}$ | $-0.27402 \times 10^{-9}$ | 491–2700 | 0.91 | 0.36 |
| Sulfur | S | 6.499 | $0.2943 \times 10^{-2}$ | $-0.1200 \times 10^{-5}$ | $0.1632 \times 10^{-9}$ | 491–3240 | 0.99 | 0.38 |
| Sulfur dioxide | $SO_2$ | 6.157 | $0.7689 \times 10^{-2}$ | $-0.2810 \times 10^{-5}$ | $0.3527 \times 10^{-9}$ | 491–3240 | 0.45 | 0.24 |
| Sulfur trioxide | $SO_3$ | 3.918 | $1.935 \times 10^{-2}$ | $-0.8256 \times 10^{-5}$ | $1.328 \times 10^{-9}$ | 491–2340 | 0.29 | 0.13 |
| Acetylene | $C_2H_2$ | 5.21 | $1.2227 \times 10^{-2}$ | $-0.4812 \times 10^{-5}$ | $0.7457 \times 10^{-9}$ | 491–2700 | 1.46 | 0.59 |
| Benzene | $C_6H_6$ | −8.650 | $6.4322 \times 10^{-2}$ | $-2.327 \times 10^{-5}$ | $3.179 \times 10^{-9}$ | 491–2700 | 0.34 | 0.20 |
| Methanol | $CH_4O$ | 4.55 | $1.214 \times 10^{-2}$ | $-0.0898 \times 10^{-5}$ | $-0.329 \times 10^{-9}$ | 491–1800 | 0.18 | 0.08 |
| Ethanol | $C_2H_6O$ | 4.75 | $2.781 \times 10^{-2}$ | $-0.7651 \times 10^{-5}$ | $0.821 \times 10^{-9}$ | 491–2700 | 0.40 | 0.22 |
| Hydrogen chloride | HCl | 7.244 | $-0.1011 \times 10^{-2}$ | $0.09783 \times 10^{-5}$ | $-0.1776 \times 10^{-9}$ | 491–2740 | 0.22 | 0.08 |
| Methane | $CH_4$ | 4.750 | $0.6666 \times 10^{-2}$ | $0.09352 \times 10^{-5}$ | $-0.4510 \times 10^{-9}$ | 491–2740 | 1.33 | 0.57 |
| Ethane | $C_2H_6$ | 1.648 | $2.291 \times 10^{-2}$ | $-0.4722 \times 10^{-5}$ | $0.2984 \times 10^{-9}$ | 491–2740 | 0.83 | 0.28 |
| Propane | $C_3H_8$ | −0.966 | $4.044 \times 10^{-2}$ | $-1.159 \times 10^{-5}$ | $1.300 \times 10^{-9}$ | 491–2740 | 0.40 | 0.12 |
| *n*-Butane | $C_4H_{10}$ | 0.945 | $4.929 \times 10^{-2}$ | $-1.352 \times 10^{-5}$ | $1.433 \times 10^{-9}$ | 491–2740 | 0.54 | 0.24 |
| *i*-Butane | $C_4H_{10}$ | −1.890 | $5.520 \times 10^{-2}$ | $-1.696 \times 10^{-5}$ | $2.044 \times 10^{-9}$ | 491–2740 | 0.25 | 0.13 |
| *n*-Pentane | $C_5H_{12}$ | 1.618 | $6.028 \times 10^{-2}$ | $-1.656 \times 10^{-5}$ | $1.732 \times 10^{-9}$ | 491–2740 | 0.56 | 0.21 |
| *n*-Hexane | $C_6H_{14}$ | 1.657 | $7.328 \times 10^{-2}$ | $-2.112 \times 10^{-5}$ | $2.363 \times 10^{-9}$ | 491–2740 | 0.72 | 0.20 |
| Ethylene | $C_2H_4$ | 0.944 | $2.075 \times 10^{-2}$ | $-0.6151 \times 10^{-5}$ | $0.7326 \times 10^{-9}$ | 491–2740 | 0.54 | 0.13 |
| Propylene | $C_3H_6$ | 0.753 | $3.162 \times 10^{-2}$ | $-0.8981 \times 10^{-5}$ | $1.008 \times 10^{-9}$ | 491–2740 | 0.73 | 0.17 |

*Source of Data:* B.G. Kyle, *Chemical and Process Thermodynamics,* 3rd ed. (Upper Saddle River, NJ: Prentice Hall, 2000).

Properties of common liquids, solids, and foods
(a) Liquids

| Substance | Boiling data at 1 atm | | Freezing data | | Liquid properties | | |
|---|---|---|---|---|---|---|---|
| | Normal boiling point, °F | Latent heat of vaporization, $h_{fg}$ Btu/lbm | Freezing point, °F | Latent heat of fusion, $h_{if}$ Btu/lbm | Temperature, °F | Density, $\rho$ lbm/ft³ | Specific heat, $c_p$ Btu/lbm·R |
| Ammonia | −27.9 | 24.54 | −107.9 | 138.6 | −27.9 | 42.6 | 1.06 |
| | | | | | 0 | 41.3 | 1.083 |
| | | | | | 40 | 39.5 | 1.103 |
| | | | | | 80 | 37.5 | 1.135 |
| Argon | −302.6 | 69.5 | −308.7 | 12.0 | −302.6 | 87.0 | 0.272 |
| Benzene | 176.4 | 169.4 | 41.9 | 54.2 | 68 | 54.9 | 0.411 |
| Brine (20% sodium chloride by mass) | 219.0 | — | 0.7 | — | 68 | 71.8 | 0.743 |
| n-Butane | 31.1 | 165.6 | −217.3 | 34.5 | 31.1 | 37.5 | 0.552 |
| Carbon dioxide | −109.2* | 99.6 (at 32°F) | −69.8 | — | 32 | 57.8 | 0.583 |
| Ethanol | 172.8 | 360.5 | −173.6 | 46.9 | 77 | 48.9 | 0.588 |
| Ethyl alcohol | 173.5 | 368 | −248.8 | 46.4 | 68 | 49.3 | 0.678 |
| Ethylene glycol | 388.6 | 344.0 | 12.6 | 77.9 | 68 | 69.2 | 0.678 |
| Glycerine | 355.8 | 419 | 66.0 | 86.3 | 68 | 78.7 | 0.554 |
| Helium | −452.1 | 9.80 | — | — | −452.1 | 9.13 | 5.45 |
| Hydrogen | −423.0 | 191.7 | −434.5 | 25.6 | −423.0 | 4.41 | 2.39 |
| Isobutane | 10.9 | 157.8 | −255.5 | 45.5 | 10.9 | 37.1 | 0.545 |
| Kerosene | 399–559 | 108 | −12.8 | — | 68 | 51.2 | 0.478 |
| Mercury | 674.1 | 126.7 | −38.0 | 4.90 | 77 | 847 | 0.033 |
| Methane | −258.7 | 219.6 | 296.0 | 25.1 | −258.7 | 26.4 | 0.834 |
| | | | | | −160 | 20.0 | 1.074 |
| Methanol | 148.1 | 473 | −143.9 | 42.7 | 77 | 49.1 | 0.609 |
| Nitrogen | −320.4 | 85.4 | −346.0 | 10.9 | −320.4 | 50.5 | 0.492 |
| | | | | | −260 | 38.2 | 0.643 |
| Octane | 256.6 | 131.7 | −71.5 | 77.9 | 68 | 43.9 | 0.502 |
| Oil (light) | — | — | | | 77 | 56.8 | 0.430 |
| Oxygen | −297.3 | 91.5 | −361.8 | 5.9 | −297.3 | 71.2 | 0.408 |
| Petroleum | — | 99–165 | | | 68 | 40.0 | 0.478 |
| Propane | −43.7 | 184.0 | −305.8 | 34.4 | −43.7 | 36.3 | 0.538 |
| | | | | | 32 | 33.0 | 0.604 |
| | | | | | 100 | 29.4 | 0.673 |
| Refrigerant-134a | −15.0 | 93.3 | −141.9 | — | −40 | 88.5 | 0.283 |
| | | | | | −15 | 86.0 | 0.294 |
| | | | | | 32 | 80.9 | 0.318 |
| | | | | | 90 | 73.6 | 0.348 |
| Water | 212 | 970.1 | 32 | 143.5 | 32 | 62.4 | 1.01 |
| | | | | | 90 | 62.1 | 1.00 |
| | | | | | 150 | 61.2 | 1.00 |
| | | | | | 212 | 59.8 | 1.01 |

*Sublimation temperature. (At pressures below the triple-point pressure of 75.1 psia, carbon dioxide exists as a solid or gas. Also, the freezing-point temperature of carbon dioxide is the triple-point temperature of −69.8°F.)

# TABLE A–3E

Properties of common liquids, solids, and foods (*Concluded*)

(*b*) Solids (values are for room temperature unless indicated otherwise)

| Substance | Density, $\rho$ lbm/ft$^3$ | Specific heat, $c_p$ Btu/lbm·R | Substance | Density, $\rho$ lbm/ft$^3$ | Specific heat, $c_p$ Btu/lbm·R |
|---|---|---|---|---|---|
| **Metals** | | | **Nonmetals** | | |
| Aluminum | | | Asphalt | 132 | 0.220 |
| −100°F | | 0.192 | Brick, common | 120 | 0.189 |
| 32°F | | 0.212 | Brick, fireclay (500°C) | 144 | 0.229 |
| 100°F | 170 | 0.218 | Concrete | 144 | 0.156 |
| 200°F | | 0.224 | Clay | 62.4 | 0.220 |
| 300°F | | 0.229 | Diamond | 151 | 0.147 |
| 400°F | | 0.235 | Glass, window | 169 | 0.191 |
| 500°F | | 0.240 | Glass, pyrex | 139 | 0.200 |
| Bronze (76% Cu, 2% Zn, 2% Al) | 517 | 0.0955 | Graphite | 156 | 0.170 |
| | | | Granite | 169 | 0.243 |
| | | | Gypsum or plaster board | | |
| Brass, yellow (65% Cu, 35% Zn) | 519 | 0.0955 | Ice | 50 | 0.260 |
| Copper | | | −50°F | | 0.424 |
| −60°F | | 0.0862 | 0°F | | 0.471 |
| 0°F | | 0.0893 | 20°F | | 0.491 |
| 100°F | 555 | 0.0925 | 32°F | 57.5 | 0.502 |
| 200°F | | 0.0938 | Limestone | 103 | 0.217 |
| 390°F | | 0.0963 | Marble | 162 | 0.210 |
| Iron | 490 | 0.107 | Plywood (Douglas fir) | 34.0 | |
| Lead | 705 | 0.030 | Rubber (hard) | 68.7 | |
| Magnesium | 108 | 0.239 | Rubber (soft) | 71.8 | |
| Nickel | 555 | 0.105 | Sand | 94.9 | |
| Silver | 655 | 0.056 | Stone | 93.6 | |
| Steel, mild | 489 | 0.119 | Woods, hard (maple, oak, etc.) | 45.0 | |
| Tungsten | 1211 | 0.031 | Woods, soft (fir, pine, etc.) | 32.0 | |

(*c*) Foods

| Food | Water content, % (mass) | Freezing point, °F | Specific heat, Btu/lbm·R Above freezing | Below freezing | Latent heat of fusion, Btu/lbm | Food | Water content, % (mass) | Freezing point, °F | Specific heat, Btu/lbm·R Above freezing | Below freezing | Latent heat of fusion, Btu/lbm |
|---|---|---|---|---|---|---|---|---|---|---|---|
| Apples | 84 | 30 | 0.873 | 0.453 | 121 | Lettuce | 95 | 32 | 0.961 | 0.487 | 136 |
| Bananas | 75 | 31 | 0.801 | 0.426 | 108 | Milk, whole | 88 | 31 | 0.905 | 0.465 | 126 |
| Beef round | 67 | — | 0.737 | 0.402 | 96 | Oranges | 87 | 31 | 0.897 | 0.462 | 125 |
| Broccoli | 90 | 31 | 0.921 | 0.471 | 129 | Potatoes | 78 | 31 | 0.825 | 0.435 | 112 |
| Butter | 16 | — | — | 0.249 | 23 | Salmon fish | 64 | 28 | 0.713 | 0.393 | 92 |
| Cheese, Swiss | 39 | 14 | 0.513 | 0.318 | 56 | Shrimp | 83 | 28 | 0.865 | 0.450 | 119 |
| Cherries | 80 | 29 | 0.841 | 0.441 | 115 | Spinach | 93 | 31 | 0.945 | 0.481 | 134 |
| Chicken | 74 | 27 | 0.793 | 0.423 | 106 | Strawberries | 90 | 31 | 0.921 | 0.471 | 129 |
| Corn, sweet | 74 | 31 | 0.793 | 0.423 | 106 | Tomatoes, ripe | 94 | 31 | 0.953 | 0.484 | 135 |
| Eggs, whole | 74 | 31 | 0.793 | 0.423 | 106 | Turkey | 64 | — | 0.713 | 0.393 | 92 |
| Ice cream | 63 | 22 | 0.705 | 0.390 | 90 | Watermelon | 93 | 31 | 0.945 | 0.481 | 134 |

*Source of Data:* Values are obtained from various handbooks and other sources or are calculated. Water content and freezing-point data of foods are from ASHRAE, *Handbook of Fundamentals,* I-P version (Atlanta, GA: American Society of Heating, Refrigerating, and Air-Conditioning Engineers, Inc., 1993), Chap. 30, Table 1. Freezing point is the temperature at which freezing starts for fruits and vegetables, and the average freezing temperature for other foods.

## TABLE A–4E

Saturated water—Temperature table

| Temp., $T\,°F$ | Sat. press., $P_{sat}$ psia | Specific volume, ft³/lbm | | Internal energy, Btu/lbm | | | Enthalpy, Btu/lbm | | | Entropy, Btu/lbm·R | | |
|---|---|---|---|---|---|---|---|---|---|---|---|---|
| | | Sat. liquid, $v_f$ | Sat. vapor, $v_g$ | Sat. liquid, $u_f$ | Evap., $u_{fg}$ | Sat. vapor, $u_g$ | Sat. liquid, $h_f$ | Evap., $h_{fg}$ | Sat. vapor, $h_g$ | Sat. liquid, $s_f$ | Evap., $s_{fg}$ | Sat. vapor, $s_g$ |
| 32.018 | 0.08871 | 0.01602 | 3299.9 | 0.000 | 1021.0 | 1021.0 | 0.000 | 1075.2 | 1075.2 | 0.00000 | 2.18672 | 2.1867 |
| 35 | 0.09998 | 0.01602 | 2945.7 | 3.004 | 1019.0 | 1022.0 | 3.004 | 1073.5 | 1076.5 | 0.00609 | 2.17011 | 2.1762 |
| 40 | 0.12173 | 0.01602 | 2443.6 | 8.032 | 1015.6 | 1023.7 | 8.032 | 1070.7 | 1078.7 | 0.01620 | 2.14271 | 2.1589 |
| 45 | 0.14756 | 0.01602 | 2035.8 | 13.05 | 1012.2 | 1025.3 | 13.05 | 1067.8 | 1080.9 | 0.02620 | 2.11587 | 2.1421 |
| 50 | 0.17812 | 0.01602 | 1703.1 | 18.07 | 1008.9 | 1026.9 | 18.07 | 1065.0 | 1083.1 | 0.03609 | 2.08956 | 2.1256 |
| 55 | 0.21413 | 0.01603 | 1430.4 | 23.07 | 1005.5 | 1028.6 | 23.07 | 1062.2 | 1085.3 | 0.04586 | 2.06377 | 2.1096 |
| 60 | 0.25638 | 0.01604 | 1206.1 | 28.08 | 1002.1 | 1030.2 | 28.08 | 1059.4 | 1087.4 | 0.05554 | 2.03847 | 2.0940 |
| 65 | 0.30578 | 0.01604 | 1020.8 | 33.08 | 998.76 | 1031.8 | 33.08 | 1056.5 | 1089.6 | 0.06511 | 2.01366 | 2.0788 |
| 70 | 0.36334 | 0.01605 | 867.18 | 38.08 | 995.39 | 1033.5 | 38.08 | 1053.7 | 1091.8 | 0.07459 | 1.98931 | 2.0639 |
| 75 | 0.43016 | 0.01606 | 739.27 | 43.07 | 992.02 | 1035.1 | 43.07 | 1050.9 | 1093.9 | 0.08398 | 1.96541 | 2.0494 |
| 80 | 0.50745 | 0.01607 | 632.41 | 48.06 | 988.65 | 1036.7 | 48.07 | 1048.0 | 1096.1 | 0.09328 | 1.94196 | 2.0352 |
| 85 | 0.59659 | 0.01609 | 542.80 | 53.06 | 985.28 | 1038.3 | 53.06 | 1045.2 | 1098.3 | 0.10248 | 1.91892 | 2.0214 |
| 90 | 0.69904 | 0.01610 | 467.40 | 58.05 | 981.90 | 1040.0 | 58.05 | 1042.4 | 1100.4 | 0.11161 | 1.89630 | 2.0079 |
| 95 | 0.81643 | 0.01612 | 403.74 | 63.04 | 978.52 | 1041.6 | 63.04 | 1039.5 | 1102.6 | 0.12065 | 1.87408 | 1.9947 |
| 100 | 0.95052 | 0.01613 | 349.83 | 68.03 | 975.14 | 1043.2 | 68.03 | 1036.7 | 1104.7 | 0.12961 | 1.85225 | 1.9819 |
| 110 | 1.2767 | 0.01617 | 264.96 | 78.01 | 968.36 | 1046.4 | 78.02 | 1031.0 | 1109.0 | 0.14728 | 1.80970 | 1.9570 |
| 120 | 1.6951 | 0.01620 | 202.94 | 88.00 | 961.56 | 1049.6 | 88.00 | 1025.2 | 1113.2 | 0.16466 | 1.76856 | 1.9332 |
| 130 | 2.2260 | 0.01625 | 157.09 | 97.99 | 954.73 | 1052.7 | 97.99 | 1019.4 | 1117.4 | 0.18174 | 1.72877 | 1.9105 |
| 140 | 2.8931 | 0.01629 | 122.81 | 107.98 | 947.87 | 1055.9 | 107.99 | 1013.6 | 1121.6 | 0.19855 | 1.69024 | 1.8888 |
| 150 | 3.7234 | 0.01634 | 96.929 | 117.98 | 940.98 | 1059.0 | 117.99 | 1007.8 | 1125.7 | 0.21508 | 1.65291 | 1.8680 |
| 160 | 4.7474 | 0.01639 | 77.185 | 127.98 | 934.05 | 1062.0 | 128.00 | 1001.8 | 1129.8 | 0.23136 | 1.61670 | 1.8481 |
| 170 | 5.9999 | 0.01645 | 61.982 | 138.00 | 927.08 | 1065.1 | 138.02 | 995.88 | 1133.9 | 0.24739 | 1.58155 | 1.8289 |
| 180 | 7.5197 | 0.01651 | 50.172 | 148.02 | 920.06 | 1068.1 | 148.04 | 989.85 | 1137.9 | 0.26318 | 1.54741 | 1.8106 |
| 190 | 9.3497 | 0.01657 | 40.920 | 158.05 | 912.99 | 1071.0 | 158.08 | 983.76 | 1141.8 | 0.27874 | 1.51421 | 1.7930 |
| 200 | 11.538 | 0.01663 | 33.613 | 168.10 | 905.87 | 1074.0 | 168.13 | 977.60 | 1145.7 | 0.29409 | 1.48191 | 1.7760 |
| 210 | 14.136 | 0.01670 | 27.798 | 178.15 | 898.68 | 1076.8 | 178.20 | 971.35 | 1149.5 | 0.30922 | 1.45046 | 1.7597 |
| 212 | 14.709 | 0.01671 | 26.782 | 180.16 | 897.24 | 1077.4 | 180.21 | 970.09 | 1150.3 | 0.31222 | 1.44427 | 1.7565 |
| 220 | 17.201 | 0.01677 | 23.136 | 188.22 | 891.43 | 1079.6 | 188.28 | 965.02 | 1153.3 | 0.32414 | 1.41980 | 1.7439 |
| 230 | 20.795 | 0.01684 | 19.374 | 198.31 | 884.10 | 1082.4 | 198.37 | 958.59 | 1157.0 | 0.33887 | 1.38989 | 1.7288 |
| 240 | 24.985 | 0.01692 | 16.316 | 208.41 | 876.70 | 1085.1 | 208.49 | 952.06 | 1160.5 | 0.35342 | 1.36069 | 1.7141 |
| 250 | 29.844 | 0.01700 | 13.816 | 218.54 | 869.21 | 1087.7 | 218.63 | 945.41 | 1164.0 | 0.36779 | 1.33216 | 1.6999 |
| 260 | 35.447 | 0.01708 | 11.760 | 228.68 | 861.62 | 1090.3 | 228.79 | 938.65 | 1167.4 | 0.38198 | 1.30425 | 1.6862 |
| 270 | 41.877 | 0.01717 | 10.059 | 238.85 | 853.94 | 1092.8 | 238.98 | 931.76 | 1170.7 | 0.39601 | 1.27694 | 1.6730 |
| 280 | 49.222 | 0.01726 | 8.6439 | 249.04 | 846.16 | 1095.2 | 249.20 | 924.74 | 1173.9 | 0.40989 | 1.25018 | 1.6601 |
| 290 | 57.573 | 0.01735 | 7.4607 | 259.26 | 838.27 | 1097.5 | 259.45 | 917.57 | 1177.0 | 0.42361 | 1.22393 | 1.6475 |
| 300 | 67.028 | 0.01745 | 6.4663 | 269.51 | 830.25 | 1099.8 | 269.73 | 910.24 | 1180.0 | 0.43720 | 1.19818 | 1.6354 |
| 310 | 77.691 | 0.01755 | 5.6266 | 279.79 | 822.11 | 1101.9 | 280.05 | 902.75 | 1182.8 | 0.45065 | 1.17289 | 1.6235 |
| 320 | 89.667 | 0.01765 | 4.9144 | 290.11 | 813.84 | 1104.0 | 290.40 | 895.09 | 1185.5 | 0.46396 | 1.14802 | 1.6120 |
| 330 | 103.07 | 0.01776 | 4.3076 | 300.46 | 805.43 | 1105.9 | 300.80 | 887.25 | 1188.1 | 0.47716 | 1.12355 | 1.6007 |
| 340 | 118.02 | 0.01787 | 3.7885 | 310.85 | 796.87 | 1107.7 | 311.24 | 879.22 | 1190.5 | 0.49024 | 1.09945 | 1.5897 |
| 350 | 134.63 | 0.01799 | 3.3425 | 321.29 | 788.16 | 1109.4 | 321.73 | 870.98 | 1192.7 | 0.50321 | 1.07570 | 1.5789 |
| 360 | 153.03 | 0.01811 | 2.9580 | 331.76 | 779.28 | 1111.0 | 332.28 | 862.53 | 1194.8 | 0.51607 | 1.05227 | 1.5683 |
| 370 | 173.36 | 0.01823 | 2.6252 | 342.29 | 770.23 | 1112.5 | 342.88 | 853.86 | 1196.7 | 0.52884 | 1.02914 | 1.5580 |
| 380 | 195.74 | 0.01836 | 2.3361 | 352.87 | 761.00 | 1113.9 | 353.53 | 844.96 | 1198.5 | 0.54152 | 1.00628 | 1.5478 |
| 390 | 220.33 | 0.01850 | 2.0842 | 363.50 | 751.58 | 1115.1 | 364.25 | 835.81 | 1200.1 | 0.55411 | 0.98366 | 1.5378 |

## TABLE A–4E

Saturated water—Temperature table *(Concluded)*

| Temp., $T\,°F$ | Sat. press., $P_{sat}$ psia | Specific volume, ft³/lbm | | Internal energy, Btu/lbm | | | Enthalpy, Btu/lbm | | | Entropy, Btu/lbm·R | | |
|---|---|---|---|---|---|---|---|---|---|---|---|---|
| | | Sat. liquid, $v_f$ | Sat. vapor, $v_g$ | Sat. liquid, $u_f$ | Evap., $u_{fg}$ | Sat. vapor, $u_g$ | Sat. liquid, $h_f$ | Evap., $h_{fg}$ | Sat. vapor, $h_g$ | Sat. liquid, $s_f$ | Evap., $s_{fg}$ | Sat. vapor, $s_g$ |
| 400 | 247.26 | 0.01864 | 1.8639 | 374.19 | 741.97 | 1116.2 | 375.04 | 826.39 | 1201.4 | 0.56663 | 0.96127 | 1.5279 |
| 410 | 276.69 | 0.01878 | 1.6706 | 384.94 | 732.14 | 1117.1 | 385.90 | 816.71 | 1202.6 | 0.57907 | 0.93908 | 1.5182 |
| 420 | 308.76 | 0.01894 | 1.5006 | 395.76 | 722.08 | 1117.8 | 396.84 | 806.74 | 1203.6 | 0.59145 | 0.91707 | 1.5085 |
| 430 | 343.64 | 0.01910 | 1.3505 | 406.65 | 711.80 | 1118.4 | 407.86 | 796.46 | 1204.3 | 0.60377 | 0.89522 | 1.4990 |
| 440 | 381.49 | 0.01926 | 1.2178 | 417.61 | 701.26 | 1118.9 | 418.97 | 785.87 | 1204.8 | 0.61603 | 0.87349 | 1.4895 |
| 450 | 422.47 | 0.01944 | 1.0999 | 428.66 | 690.47 | 1119.1 | 430.18 | 774.94 | 1205.1 | 0.62826 | 0.85187 | 1.4801 |
| 460 | 466.75 | 0.01962 | 0.99510 | 439.79 | 679.39 | 1119.2 | 441.48 | 763.65 | 1205.1 | 0.64044 | 0.83033 | 1.4708 |
| 470 | 514.52 | 0.01981 | 0.90158 | 451.01 | 668.02 | 1119.0 | 452.90 | 751.98 | 1204.9 | 0.65260 | 0.80885 | 1.4615 |
| 480 | 565.96 | 0.02001 | 0.81794 | 462.34 | 656.34 | 1118.7 | 464.43 | 739.91 | 1204.3 | 0.66474 | 0.78739 | 1.4521 |
| 490 | 621.24 | 0.02022 | 0.74296 | 473.77 | 644.32 | 1118.1 | 476.09 | 727.40 | 1203.5 | 0.67686 | 0.76594 | 1.4428 |
| 500 | 680.56 | 0.02044 | 0.67558 | 485.32 | 631.94 | 1117.3 | 487.89 | 714.44 | 1202.3 | 0.68899 | 0.74445 | 1.4334 |
| 510 | 744.11 | 0.02067 | 0.61489 | 496.99 | 619.17 | 1116.2 | 499.84 | 700.99 | 1200.8 | 0.70112 | 0.72290 | 1.4240 |
| 520 | 812.11 | 0.02092 | 0.56009 | 508.80 | 605.99 | 1114.8 | 511.94 | 687.01 | 1199.0 | 0.71327 | 0.70126 | 1.4145 |
| 530 | 884.74 | 0.02118 | 0.51051 | 520.76 | 592.35 | 1113.1 | 524.23 | 672.47 | 1196.7 | 0.72546 | 0.67947 | 1.4049 |
| 540 | 962.24 | 0.02146 | 0.46553 | 532.88 | 578.23 | 1111.1 | 536.70 | 657.31 | 1194.0 | 0.73770 | 0.65751 | 1.3952 |
| 550 | 1044.8 | 0.02176 | 0.42465 | 545.18 | 563.58 | 1108.8 | 549.39 | 641.47 | 1190.9 | 0.75000 | 0.63532 | 1.3853 |
| 560 | 1132.7 | 0.02207 | 0.38740 | 557.68 | 548.33 | 1106.0 | 562.31 | 624.91 | 1187.2 | 0.76238 | 0.61284 | 1.3752 |
| 570 | 1226.2 | 0.02242 | 0.35339 | 570.40 | 532.45 | 1102.8 | 575.49 | 607.55 | 1183.0 | 0.77486 | 0.59003 | 1.3649 |
| 580 | 1325.5 | 0.02279 | 0.32225 | 583.37 | 515.84 | 1099.2 | 588.95 | 589.29 | 1178.2 | 0.78748 | 0.56679 | 1.3543 |
| 590 | 1430.8 | 0.02319 | 0.29367 | 596.61 | 498.43 | 1095.0 | 602.75 | 570.04 | 1172.8 | 0.80026 | 0.54306 | 1.3433 |
| 600 | 1542.5 | 0.02362 | 0.26737 | 610.18 | 480.10 | 1090.3 | 616.92 | 549.67 | 1166.6 | 0.81323 | 0.51871 | 1.3319 |
| 610 | 1660.9 | 0.02411 | 0.24309 | 624.11 | 460.73 | 1084.8 | 631.52 | 528.03 | 1159.5 | 0.82645 | 0.49363 | 1.3201 |
| 620 | 1786.2 | 0.02464 | 0.22061 | 638.47 | 440.14 | 1078.6 | 646.62 | 504.92 | 1151.5 | 0.83998 | 0.46765 | 1.3076 |
| 630 | 1918.9 | 0.02524 | 0.19972 | 653.35 | 418.12 | 1071.5 | 662.32 | 480.07 | 1142.4 | 0.85389 | 0.44056 | 1.2944 |
| 640 | 2059.3 | 0.02593 | 0.18019 | 668.86 | 394.36 | 1063.2 | 678.74 | 453.14 | 1131.9 | 0.86828 | 0.41206 | 1.2803 |
| 650 | 2207.8 | 0.02673 | 0.16184 | 685.16 | 368.44 | 1053.6 | 696.08 | 423.65 | 1119.7 | 0.88332 | 0.38177 | 1.2651 |
| 660 | 2364.9 | 0.02767 | 0.14444 | 702.48 | 339.74 | 1042.2 | 714.59 | 390.84 | 1105.4 | 0.89922 | 0.34906 | 1.2483 |
| 670 | 2531.2 | 0.02884 | 0.12774 | 721.23 | 307.22 | 1028.5 | 734.74 | 353.54 | 1088.3 | 0.91636 | 0.31296 | 1.2293 |
| 680 | 2707.3 | 0.03035 | 0.11134 | 742.11 | 269.00 | 1011.1 | 757.32 | 309.57 | 1066.9 | 0.93541 | 0.27163 | 1.2070 |
| 690 | 2894.1 | 0.03255 | 0.09451 | 766.81 | 220.77 | 987.6 | 784.24 | 253.96 | 1038.2 | 0.95797 | 0.22089 | 1.1789 |
| 700 | 3093.0 | 0.03670 | 0.07482 | 801.75 | 146.50 | 948.3 | 822.76 | 168.32 | 991.1 | 0.99023 | 0.14514 | 1.1354 |
| 705.10 | 3200.1 | 0.04975 | 0.04975 | 866.61 | 0 | 866.6 | 896.07 | 0 | 896.1 | 1.05257 | 0 | 1.0526 |

*Source of Data:* Tables A–4E through A–8E are generated using the Engineering Equation Solver (EES) software developed by S. A. Klein and F. L. Alvarado. The routine used in calculations is the highly accurate Steam_IAPWS, which incorporates the 1995 Formulation for the Thermodynamic Properties of Ordinary Water Substance for General and Scientific Use, issued by The International Association for the Properties of Water and Steam (IAPWS). This formulation replaces the 1984 formulation of Haar, Gallagher, and Kell *(NBS/NRC Steam Tables,* Hemisphere Publishing Co., 1984), which is also available in EES as the routine STEAM. The new formulation is based on the correlations of Saul and Wagner *(J. Phys. Chem. Ref. Data,* 16, 893, 1987) with modifications to adjust to the International Temperature Scale of 1990. The modifications are described by Wagner and Pruss *(J. Phys. Chem. Ref. Data,* 22, 783, 1993). The properties of ice are based on Hyland and Wexler, "Formulations for the Thermodynamic Properties of the Saturated Phases of $H_2O$ from 173.15 K to 473.15 K," *ASHRAE Trans.,* Part 2A, Paper 2793, 1983.

Saturated water—Pressure table

| Press., P psia | Sat. temp., $T_{sat}$ °F | Specific volume, ft³/lbm | | Internal energy, Btu/lbm | | | Enthalpy, Btu/lbm | | | Entropy, Btu/lbm·R | | |
|---|---|---|---|---|---|---|---|---|---|---|---|---|
| | | Sat. liquid, $v_f$ | Sat. vapor, $v_g$ | Sat. liquid, $u_f$ | Evap., $u_{fg}$ | Sat. vapor, $u_g$ | Sat. liquid, $h_f$ | Evap., $h_{fg}$ | Sat. vapor, $h_g$ | Sat. liquid, $s_f$ | Evap., $s_{fg}$ | Sat. vapor, $s_g$ |
| 1 | 101.69 | 0.01614 | 333.49 | 69.72 | 973.99 | 1043.7 | 69.72 | 1035.7 | 1105.4 | 0.13262 | 1.84495 | 1.9776 |
| 2 | 126.02 | 0.01623 | 173.71 | 94.02 | 957.45 | 1051.5 | 94.02 | 1021.7 | 1115.8 | 0.17499 | 1.74444 | 1.9194 |
| 3 | 141.41 | 0.01630 | 118.70 | 109.39 | 946.90 | 1056.3 | 109.40 | 1012.8 | 1122.2 | 0.20090 | 1.68489 | 1.8858 |
| 4 | 152.91 | 0.01636 | 90.629 | 120.89 | 938.97 | 1059.9 | 120.90 | 1006.0 | 1126.9 | 0.21985 | 1.64225 | 1.8621 |
| 5 | 162.18 | 0.01641 | 73.525 | 130.17 | 932.53 | 1062.7 | 130.18 | 1000.5 | 1130.7 | 0.23488 | 1.60894 | 1.8438 |
| 6 | 170.00 | 0.01645 | 61.982 | 138.00 | 927.08 | 1065.1 | 138.02 | 995.88 | 1133.9 | 0.24739 | 1.58155 | 1.8289 |
| 8 | 182.81 | 0.01652 | 47.347 | 150.83 | 918.08 | 1068.9 | 150.86 | 988.15 | 1139.0 | 0.26757 | 1.53800 | 1.8056 |
| 10 | 193.16 | 0.01659 | 38.425 | 161.22 | 910.75 | 1072.0 | 161.25 | 981.82 | 1143.1 | 0.28362 | 1.50391 | 1.7875 |
| 14.696 | 211.95 | 0.01671 | 26.805 | 180.12 | 897.27 | 1077.4 | 180.16 | 970.12 | 1150.3 | 0.31215 | 1.44441 | 1.7566 |
| 15 | 212.99 | 0.01672 | 26.297 | 181.16 | 896.52 | 1077.7 | 181.21 | 969.47 | 1150.7 | 0.31370 | 1.44441 | 1.7549 |
| 20 | 227.92 | 0.01683 | 20.093 | 196.21 | 885.63 | 1081.8 | 196.27 | 959.93 | 1156.2 | 0.33582 | 1.39606 | 1.7319 |
| 25 | 240.03 | 0.01692 | 16.307 | 208.45 | 876.67 | 1085.1 | 208.52 | 952.03 | 1160.6 | 0.35347 | 1.36060 | 1.7141 |
| 30 | 250.30 | 0.01700 | 13.749 | 218.84 | 868.98 | 1087.8 | 218.93 | 945.21 | 1164.1 | 0.36821 | 1.33132 | 1.6995 |
| 35 | 259.25 | 0.01708 | 11.901 | 227.92 | 862.19 | 1090.1 | 228.03 | 939.16 | 1167.2 | 0.38093 | 1.30632 | 1.6872 |
| 40 | 267.22 | 0.01715 | 10.501 | 236.02 | 856.09 | 1092.1 | 236.14 | 933.69 | 1169.8 | 0.39213 | 1.28448 | 1.6766 |
| 45 | 274.41 | 0.01721 | 9.4028 | 243.34 | 850.52 | 1093.9 | 243.49 | 928.68 | 1172.2 | 0.40216 | 1.26506 | 1.6672 |
| 50 | 280.99 | 0.01727 | 8.5175 | 250.05 | 845.39 | 1095.4 | 250.21 | 924.03 | 1174.2 | 0.41125 | 1.24756 | 1.6588 |
| 55 | 287.05 | 0.01732 | 7.7882 | 256.25 | 840.61 | 1096.9 | 256.42 | 919.70 | 1176.1 | 0.41958 | 1.23162 | 1.6512 |
| 60 | 292.69 | 0.01738 | 7.1766 | 262.01 | 836.13 | 1098.1 | 262.20 | 915.61 | 1177.8 | 0.42728 | 1.21697 | 1.6442 |
| 65 | 297.95 | 0.01743 | 6.6560 | 267.41 | 831.90 | 1099.3 | 267.62 | 911.75 | 1179.4 | 0.43443 | 1.20341 | 1.6378 |
| 70 | 302.91 | 0.01748 | 6.2075 | 272.50 | 827.90 | 1100.4 | 272.72 | 908.08 | 1180.8 | 0.44112 | 1.19078 | 1.6319 |
| 75 | 307.59 | 0.01752 | 5.8167 | 277.31 | 824.09 | 1101.4 | 277.55 | 904.58 | 1182.1 | 0.44741 | 1.17895 | 1.6264 |
| 80 | 312.02 | 0.01757 | 5.4733 | 281.87 | 820.45 | 1102.3 | 282.13 | 901.22 | 1183.4 | 0.45335 | 1.16783 | 1.6212 |
| 85 | 316.24 | 0.01761 | 5.1689 | 286.22 | 816.97 | 1103.2 | 286.50 | 898.00 | 1184.5 | 0.45897 | 1.15732 | 1.6163 |
| 90 | 320.26 | 0.01765 | 4.8972 | 290.38 | 813.62 | 1104.0 | 290.67 | 894.89 | 1185.6 | 0.46431 | 1.14737 | 1.6117 |
| 95 | 324.11 | 0.01770 | 4.6532 | 294.36 | 810.40 | 1104.8 | 294.67 | 891.89 | 1186.6 | 0.46941 | 1.13791 | 1.6073 |
| 100 | 327.81 | 0.01774 | 4.4327 | 298.19 | 807.29 | 1105.5 | 298.51 | 888.99 | 1187.5 | 0.47427 | 1.12888 | 1.6032 |
| 110 | 334.77 | 0.01781 | 4.0410 | 305.41 | 801.37 | 1106.8 | 305.78 | 883.44 | 1189.2 | 0.48341 | 1.11201 | 1.5954 |
| 120 | 341.25 | 0.01789 | 3.7289 | 312.16 | 795.79 | 1107.9 | 312.55 | 878.20 | 1190.8 | 0.49187 | 1.09646 | 1.5883 |
| 130 | 347.32 | 0.01796 | 3.4557 | 318.48 | 790.51 | 1109.0 | 318.92 | 873.21 | 1192.1 | 0.49974 | 1.08204 | 1.5818 |
| 140 | 353.03 | 0.01802 | 3.2202 | 324.45 | 785.49 | 1109.9 | 324.92 | 868.45 | 1193.4 | 0.50711 | 1.06858 | 1.5757 |
| 150 | 358.42 | 0.01809 | 3.0150 | 330.11 | 780.69 | 1110.8 | 330.61 | 863.88 | 1194.5 | 0.51405 | 1.05595 | 1.5700 |
| 160 | 363.54 | 0.01815 | 2.8347 | 335.49 | 776.10 | 1111.6 | 336.02 | 859.49 | 1195.5 | 0.52061 | 1.04405 | 1.5647 |
| 170 | 368.41 | 0.01821 | 2.6749 | 340.62 | 771.68 | 1112.3 | 341.19 | 855.25 | 1196.4 | 0.52682 | 1.03279 | 1.5596 |
| 180 | 373.07 | 0.01827 | 2.5322 | 345.53 | 767.42 | 1113.0 | 346.14 | 851.16 | 1197.3 | 0.53274 | 1.02210 | 1.5548 |
| 190 | 377.52 | 0.01833 | 2.4040 | 350.24 | 763.31 | 1113.6 | 350.89 | 847.19 | 1198.1 | 0.53839 | 1.01191 | 1.5503 |
| 200 | 381.80 | 0.01839 | 2.2882 | 354.78 | 759.32 | 1114.1 | 355.46 | 843.33 | 1198.8 | 0.54379 | 1.00219 | 1.5460 |
| 250 | 400.97 | 0.01865 | 1.8440 | 375.23 | 741.02 | 1116.3 | 376.09 | 825.47 | 1201.6 | 0.56784 | 0.95912 | 1.5270 |
| 300 | 417.35 | 0.01890 | 1.5435 | 392.89 | 724.77 | 1117.7 | 393.94 | 809.41 | 1203.3 | 0.58818 | 0.92289 | 1.5111 |
| 350 | 431.74 | 0.01912 | 1.3263 | 408.55 | 709.98 | 1118.5 | 409.79 | 794.65 | 1204.4 | 0.60590 | 0.89143 | 1.4973 |
| 400 | 444.62 | 0.01934 | 1.1617 | 422.70 | 696.31 | 1119.0 | 424.13 | 780.87 | 1205.0 | 0.62168 | 0.86350 | 1.4852 |
| 450 | 456.31 | 0.01955 | 1.0324 | 435.67 | 683.52 | 1119.2 | 437.30 | 767.86 | 1205.2 | 0.63595 | 0.83828 | 1.4742 |
| 500 | 467.04 | 0.01975 | 0.92819 | 447.68 | 671.42 | 1119.1 | 449.51 | 755.48 | 1205.0 | 0.64900 | 0.81521 | 1.4642 |
| 550 | 476.97 | 0.01995 | 0.84228 | 458.90 | 659.91 | 1118.8 | 460.93 | 743.60 | 1204.5 | 0.66107 | 0.79388 | 1.4550 |
| 600 | 486.24 | 0.02014 | 0.77020 | 469.46 | 648.88 | 1118.3 | 471.70 | 732.15 | 1203.9 | 0.67231 | 0.77400 | 1.4463 |

Saturated water—Pressure table *(Concluded)*

| Press., $P$ psia | Sat. temp., $T_{sat}$ °F | Specific volume, ft³/lbm | | Internal energy, Btu/lbm | | | Enthalpy, Btu/lbm | | | Entropy, Btu/lbm·R | | |
|---|---|---|---|---|---|---|---|---|---|---|---|---|
| | | Sat. liquid, $v_f$ | Sat. vapor, $v_g$ | Sat. liquid, $u_f$ | Evap., $u_{fg}$ | Sat. vapor, $u_g$ | Sat. liquid, $h_f$ | Evap., $h_{fg}$ | Sat. vapor, $h_g$ | Sat. liquid, $s_f$ | Evap., $s_{fg}$ | Sat. vapor, $s_g$ |
| 700 | 503.13 | 0.02051 | 0.65589 | 488.96 | 627.98 | 1116.9 | 491.62 | 710.29 | 1201.9 | 0.69279 | 0.73771 | 1.4305 |
| 800 | 518.27 | 0.02087 | 0.56920 | 506.74 | 608.30 | 1115.0 | 509.83 | 689.48 | 1199.3 | 0.71117 | 0.70502 | 1.4162 |
| 900 | 532.02 | 0.02124 | 0.50107 | 523.19 | 589.54 | 1112.7 | 526.73 | 669.46 | 1196.2 | 0.72793 | 0.67505 | 1.4030 |
| 1000 | 544.65 | 0.02159 | 0.44604 | 538.58 | 571.49 | 1110.1 | 542.57 | 650.03 | 1192.6 | 0.74341 | 0.64722 | 1.3906 |
| 1200 | 567.26 | 0.02232 | 0.36241 | 566.89 | 536.87 | 1103.8 | 571.85 | 612.39 | 1184.2 | 0.77143 | 0.59632 | 1.3677 |
| 1400 | 587.14 | 0.02307 | 0.30161 | 592.79 | 503.50 | 1096.3 | 598.76 | 575.66 | 1174.4 | 0.79658 | 0.54991 | 1.3465 |
| 1600 | 604.93 | 0.02386 | 0.25516 | 616.99 | 470.69 | 1087.7 | 624.06 | 539.18 | 1163.2 | 0.81972 | 0.50645 | 1.3262 |
| 1800 | 621.07 | 0.02470 | 0.21831 | 640.03 | 437.86 | 1077.9 | 648.26 | 502.35 | 1150.6 | 0.84144 | 0.46482 | 1.3063 |
| 2000 | 635.85 | 0.02563 | 0.18815 | 662.33 | 404.46 | 1066.8 | 671.82 | 464.60 | 1136.4 | 0.86224 | 0.42409 | 1.2863 |
| 2500 | 668.17 | 0.02860 | 0.13076 | 717.67 | 313.53 | 1031.2 | 730.90 | 360.79 | 1091.7 | 0.91311 | 0.31988 | 1.2330 |
| 3000 | 695.41 | 0.03433 | 0.08460 | 783.39 | 186.41 | 969.8 | 802.45 | 214.32 | 1016.8 | 0.97321 | 0.18554 | 1.1587 |
| 3200.1 | 705.10 | 0.04975 | 0.04975 | 866.61 | 0 | 866.6 | 896.07 | 0 | 896.1 | 1.05257 | 0 | 1.0526 |

## TABLE A–6E

Superheated water

| T °F | $v$ ft³/lbm | $u$ Btu/lbm | $h$ Btu/lbm | $s$ Btu/lbm·R | $v$ ft³/lbm | $u$ Btu/lbm | $h$ Btu/lbm | $s$ Btu/lbm·R | $v$ ft³/lbm | $u$ Btu/lbm | $h$ Btu/lbm | $s$ Btu/lbm·R |
|---|---|---|---|---|---|---|---|---|---|---|---|---|
| | $P = 1.0$ psia (101.69°F)* | | | | $P = 5.0$ psia (162.18°F) | | | | $P = 10$ psia (193.16°F) | | | |
| Sat.† | 333.49 | 1043.7 | 1105.4 | 1.9776 | 73.525 | 1062.7 | 1130.7 | 1.8438 | 38.425 | 1072.0 | 1143.1 | 1.7875 |
| 200 | 392.53 | 1077.5 | 1150.1 | 2.0509 | 78.153 | 1076.2 | 1148.5 | 1.8716 | 38.849 | 1074.5 | 1146.4 | 1.7926 |
| 240 | 416.44 | 1091.2 | 1168.3 | 2.0777 | 83.009 | 1090.3 | 1167.1 | 1.8989 | 41.326 | 1089.1 | 1165.5 | 1.8207 |
| 280 | 440.33 | 1105.0 | 1186.5 | 2.1030 | 87.838 | 1104.3 | 1185.6 | 1.9246 | 43.774 | 1103.4 | 1184.4 | 1.8469 |
| 320 | 464.20 | 1118.9 | 1204.8 | 2.1271 | 92.650 | 1118.4 | 1204.1 | 1.9490 | 46.205 | 1117.6 | 1203.1 | 1.8716 |
| 360 | 488.07 | 1132.9 | 1223.3 | 2.1502 | 97.452 | 1132.5 | 1222.6 | 1.9722 | 48.624 | 1131.9 | 1221.8 | 1.8950 |
| 400 | 511.92 | 1147.1 | 1241.8 | 2.1722 | 102.25 | 1146.7 | 1241.3 | 1.9944 | 51.035 | 1146.2 | 1240.6 | 1.9174 |
| 440 | 535.77 | 1161.3 | 1260.4 | 2.1934 | 107.03 | 1160.9 | 1260.0 | 2.0156 | 53.441 | 1160.5 | 1259.4 | 1.9388 |
| 500 | 571.54 | 1182.8 | 1288.6 | 2.2237 | 114.21 | 1182.6 | 1288.2 | 2.0461 | 57.041 | 1182.2 | 1287.8 | 1.9693 |
| 600 | 631.14 | 1219.4 | 1336.2 | 2.2709 | 126.15 | 1219.2 | 1335.9 | 2.0933 | 63.029 | 1219.0 | 1335.6 | 2.0167 |
| 700 | 690.73 | 1256.8 | 1384.6 | 2.3146 | 138.09 | 1256.7 | 1384.4 | 2.1371 | 69.007 | 1256.5 | 1384.2 | 2.0605 |
| 800 | 750.31 | 1295.1 | 1433.9 | 2.3553 | 150.02 | 1294.9 | 1433.7 | 2.1778 | 74.980 | 1294.8 | 1433.5 | 2.1013 |
| 1000 | 869.47 | 1374.2 | 1535.1 | 2.4299 | 173.86 | 1374.2 | 1535.0 | 2.2524 | 86.913 | 1374.1 | 1534.9 | 2.1760 |
| 1200 | 988.62 | 1457.1 | 1640.0 | 2.4972 | 197.70 | 1457.0 | 1640.0 | 2.3198 | 98.840 | 1457.0 | 1639.9 | 2.2433 |
| 1400 | 1107.8 | 1543.7 | 1748.7 | 2.5590 | 221.54 | 1543.7 | 1748.7 | 2.3816 | 110.762 | 1543.6 | 1748.6 | 2.3052 |
| | $P = 15$ psia (212.99°F) | | | | $P = 20$ psia (227.92°F) | | | | $P = 40$ psia (267.22°F) | | | |
| Sat. | 26.297 | 1077.7 | 1150.7 | 1.7549 | 20.093 | 1081.8 | 1156.2 | 1.7319 | 10.501 | 1092.1 | 1169.8 | 1.6766 |
| 240 | 27.429 | 1087.8 | 1163.9 | 1.7742 | 20.478 | 1086.5 | 1162.3 | 1.7406 | | | | |
| 280 | 29.085 | 1102.4 | 1183.2 | 1.8010 | 21.739 | 1101.4 | 1181.9 | 1.7679 | 10.713 | 1097.3 | 1176.6 | 1.6858 |
| 320 | 30.722 | 1116.9 | 1202.2 | 1.8260 | 22.980 | 1116.1 | 1201.2 | 1.7933 | 11.363 | 1112.9 | 1197.1 | 1.7128 |
| 360 | 32.348 | 1131.3 | 1221.1 | 1.8496 | 24.209 | 1130.7 | 1220.2 | 1.8171 | 11.999 | 1128.1 | 1216.9 | 1.7376 |
| 400 | 33.965 | 1145.7 | 1239.9 | 1.8721 | 25.429 | 1145.1 | 1239.3 | 1.8398 | 12.625 | 1143.1 | 1236.5 | 1.7610 |
| 440 | 35.576 | 1160.1 | 1258.8 | 1.8936 | 26.644 | 1159.7 | 1258.3 | 1.8614 | 13.244 | 1157.9 | 1256.0 | 1.7831 |
| 500 | 37.986 | 1181.9 | 1287.3 | 1.9243 | 28.458 | 1181.6 | 1286.9 | 1.8922 | 14.165 | 1180.2 | 1285.0 | 1.8143 |
| 600 | 41.988 | 1218.7 | 1335.3 | 1.9718 | 31.467 | 1218.5 | 1334.9 | 1.9398 | 15.686 | 1217.5 | 1333.6 | 1.8625 |
| 700 | 45.981 | 1256.3 | 1383.9 | 2.0156 | 34.467 | 1256.1 | 1383.7 | 1.9837 | 17.197 | 1255.3 | 1382.6 | 1.9067 |
| 800 | 49.967 | 1294.6 | 1433.3 | 2.0565 | 37.461 | 1294.5 | 1433.1 | 2.0247 | 18.702 | 1293.9 | 1432.3 | 1.9478 |
| 1000 | 57.930 | 1374.0 | 1534.8 | 2.1312 | 43.438 | 1373.8 | 1534.6 | 2.0994 | 21.700 | 1373.4 | 1534.1 | 2.0227 |
| 1200 | 65.885 | 1456.9 | 1639.8 | 2.1986 | 49.407 | 1456.8 | 1639.7 | 2.1668 | 24.691 | 1456.5 | 1639.3 | 2.0902 |
| 1400 | 73.836 | 1543.6 | 1748.5 | 2.2604 | 55.373 | 1543.5 | 1748.4 | 2.2287 | 27.678 | 1543.3 | 1748.1 | 2.1522 |
| 1600 | 81.784 | 1634.0 | 1861.0 | 2.3178 | 61.335 | 1633.9 | 1860.9 | 2.2861 | 30.662 | 1633.7 | 1860.7 | 2.2096 |
| | $P = 60$ psia (292.69°F) | | | | $P = 80$ psia (312.02°F) | | | | $P = 100$ psia (327.81°F) | | | |
| Sat. | 7.1766 | 1098.1 | 1177.8 | 1.6442 | 5.4733 | 1102.3 | 1183.4 | 1.6212 | 4.4327 | 1105.5 | 1187.5 | 1.6032 |
| 320 | 7.4863 | 1109.6 | 1192.7 | 1.6636 | 5.5440 | 1105.9 | 1187.9 | 1.6271 | | | | |
| 360 | 7.9259 | 1125.5 | 1213.5 | 1.6897 | 5.8876 | 1122.7 | 1209.9 | 1.6545 | 4.6628 | 1119.8 | 1206.1 | 1.6263 |
| 400 | 8.3548 | 1140.9 | 1233.7 | 1.7138 | 6.2187 | 1138.7 | 1230.8 | 1.6794 | 4.9359 | 1136.4 | 1227.8 | 1.6521 |
| 440 | 8.7766 | 1156.1 | 1253.6 | 1.7364 | 6.5420 | 1154.3 | 1251.2 | 1.7026 | 5.2006 | 1152.4 | 1248.7 | 1.6759 |
| 500 | 9.4005 | 1178.8 | 1283.1 | 1.7682 | 7.0177 | 1177.3 | 1281.2 | 1.7350 | 5.5876 | 1175.9 | 1279.3 | 1.7088 |
| 600 | 10.4256 | 1216.5 | 1332.2 | 1.8168 | 7.7951 | 1215.4 | 1330.8 | 1.7841 | 6.2167 | 1214.4 | 1329.4 | 1.7586 |
| 700 | 11.4401 | 1254.5 | 1381.6 | 1.8613 | 8.5616 | 1253.8 | 1380.5 | 1.8289 | 6.8344 | 1253.0 | 1379.5 | 1.8037 |
| 800 | 12.4484 | 1293.3 | 1431.5 | 1.9026 | 9.3218 | 1292.6 | 1430.6 | 1.8704 | 7.4457 | 1292.0 | 1429.8 | 1.8453 |
| 1000 | 14.4543 | 1373.0 | 1533.5 | 1.9777 | 10.8313 | 1372.6 | 1532.9 | 1.9457 | 8.6575 | 1372.2 | 1532.4 | 1.9208 |
| 1200 | 16.4525 | 1456.2 | 1638.9 | 2.0454 | 12.3331 | 1455.9 | 1638.5 | 2.0135 | 9.8615 | 1455.6 | 1638.1 | 1.9887 |
| 1400 | 18.4464 | 1543.0 | 1747.8 | 2.1073 | 13.8306 | 1542.8 | 1747.5 | 2.0755 | 11.0612 | 1542.6 | 1747.2 | 2.0508 |
| 1600 | 20.438 | 1633.5 | 1860.5 | 2.1648 | 15.3257 | 1633.3 | 1860.2 | 2.1330 | 12.2584 | 1633.2 | 1860.0 | 2.1083 |
| 1800 | 22.428 | 1727.6 | 1976.6 | 2.2187 | 16.8192 | 1727.5 | 1976.5 | 2.1869 | 13.4541 | 1727.3 | 1976.3 | 2.1622 |
| 2000 | 24.417 | 1825.2 | 2096.3 | 2.2694 | 18.3117 | 1825.0 | 2096.1 | 2.2376 | 14.6487 | 1824.9 | 2096.0 | 2.2130 |

*The temperature in parentheses is the saturation temperature at the specified pressure.
†Properties of saturated vapor at the specified pressure.

Superheated water  (Continued)

| T °F | υ ft³/lbm | u Btu/lbm | h Btu/lbm | s Btu/lbm·R | υ ft³/lbm | u Btu/lbm | h Btu/lbm | s Btu/lbm·R | υ ft³/lbm | u Btu/lbm | h Btu/lbm | s Btu/lbm·R |
|---|---|---|---|---|---|---|---|---|---|---|---|---|
| | P = 120 psia (341.25°F) | | | | P = 140 psia (353.03°F) | | | | P = 160 psia (363.54°F) | | | |
| Sat. | 3.7289 | 1107.9 | 1190.8 | 1.5883 | 3.2202 | 1109.9 | 1193.4 | 1.5757 | 2.8347 | 1111.6 | 1195.5 | 1.5647 |
| 360 | 3.8446 | 1116.7 | 1202.1 | 1.6023 | 3.2584 | 1113.4 | 1197.8 | 1.5811 | | | | |
| 400 | 4.0799 | 1134.0 | 1224.6 | 1.6292 | 3.4676 | 1131.5 | 1221.4 | 1.6092 | 3.0076 | 1129.0 | 1218.0 | 1.5914 |
| 450 | 4.3613 | 1154.5 | 1251.4 | 1.6594 | 3.7147 | 1152.6 | 1248.9 | 1.6403 | 3.2293 | 1150.7 | 1246.3 | 1.6234 |
| 500 | 4.6340 | 1174.4 | 1277.3 | 1.6872 | 3.9525 | 1172.9 | 1275.3 | 1.6686 | 3.4412 | 1171.4 | 1273.2 | 1.6522 |
| 550 | 4.9010 | 1193.9 | 1302.8 | 1.7131 | 4.1845 | 1192.7 | 1301.1 | 1.6948 | 3.6469 | 1191.4 | 1299.4 | 1.6788 |
| 600 | 5.1642 | 1213.4 | 1328.0 | 1.7375 | 4.4124 | 1212.3 | 1326.6 | 1.7195 | 3.8484 | 1211.3 | 1325.2 | 1.7037 |
| 700 | 5.6829 | 1252.2 | 1378.4 | 1.7829 | 4.8604 | 1251.4 | 1377.3 | 1.7652 | 4.2434 | 1250.6 | 1376.3 | 1.7498 |
| 800 | 6.1950 | 1291.4 | 1429.0 | 1.8247 | 5.3017 | 1290.8 | 1428.1 | 1.8072 | 4.6316 | 1290.2 | 1427.3 | 1.7920 |
| 1000 | 7.2083 | 1371.7 | 1531.8 | 1.9005 | 6.1732 | 1371.3 | 1531.3 | 1.8832 | 5.3968 | 1370.9 | 1530.7 | 1.8682 |
| 1200 | 8.2137 | 1455.3 | 1637.7 | 1.9684 | 7.0367 | 1455.0 | 1637.3 | 1.9512 | 6.1540 | 1454.7 | 1636.9 | 1.9363 |
| 1400 | 9.2149 | 1542.3 | 1746.9 | 2.0305 | 7.8961 | 1542.1 | 1746.6 | 2.0134 | 6.9070 | 1541.8 | 1746.3 | 1.9986 |
| 1600 | 10.2135 | 1633.0 | 1859.8 | 2.0881 | 8.7529 | 1632.8 | 1859.5 | 2.0711 | 7.6574 | 1632.6 | 1859.3 | 2.0563 |
| 1800 | 11.2106 | 1727.2 | 1976.1 | 2.1420 | 9.6082 | 1727.0 | 1975.9 | 2.1250 | 8.4063 | 1726.9 | 1975.7 | 2.1102 |
| 2000 | 12.2067 | 1824.8 | 2095.8 | 2.1928 | 10.4624 | 1824.6 | 2095.7 | 2.1758 | 9.1542 | 1824.5 | 2095.5 | 2.1610 |
| | P = 180 psia (373.07°F) | | | | P = 200 psia (381.80°F) | | | | P = 225 psia (391.80°F) | | | |
| Sat. | 2.5322 | 1113.0 | 1197.3 | 1.5548 | 2.2882 | 1114.1 | 1198.8 | 1.5460 | 2.0423 | 1115.3 | 1200.3 | 1.5360 |
| 400 | 2.6490 | 1126.3 | 1214.5 | 1.5752 | 2.3615 | 1123.5 | 1210.9 | 1.5602 | 2.0728 | 1119.7 | 1206.0 | 1.5427 |
| 450 | 2.8514 | 1148.7 | 1243.7 | 1.6082 | 2.5488 | 1146.7 | 1241.0 | 1.5943 | 2.2457 | 1144.1 | 1237.6 | 1.5783 |
| 500 | 3.0433 | 1169.8 | 1271.2 | 1.6376 | 2.7247 | 1168.2 | 1269.0 | 1.6243 | 2.4059 | 1166.2 | 1266.3 | 1.6091 |
| 550 | 3.2286 | 1190.2 | 1297.7 | 1.6646 | 2.8939 | 1188.9 | 1296.0 | 1.6516 | 2.5590 | 1187.2 | 1293.8 | 1.6370 |
| 600 | 3.4097 | 1210.2 | 1323.8 | 1.6897 | 3.0586 | 1209.1 | 1322.3 | 1.6771 | 2.7075 | 1207.7 | 1320.5 | 1.6628 |
| 700 | 3.7635 | 1249.8 | 1375.2 | 1.7361 | 3.3796 | 1249.0 | 1374.1 | 1.7238 | 2.9956 | 1248.0 | 1372.7 | 1.7099 |
| 800 | 4.1104 | 1289.5 | 1426.5 | 1.7785 | 3.6934 | 1288.9 | 1425.6 | 1.7664 | 3.2765 | 1288.1 | 1424.5 | 1.7528 |
| 900 | 4.4531 | 1329.7 | 1478.0 | 1.8179 | 4.0031 | 1329.2 | 1477.3 | 1.8059 | 3.5530 | 1328.5 | 1476.5 | 1.7925 |
| 1000 | 4.7929 | 1370.5 | 1530.1 | 1.8549 | 4.3099 | 1370.1 | 1529.6 | 1.8430 | 3.8268 | 1369.5 | 1528.9 | 1.8296 |
| 1200 | 5.4674 | 1454.3 | 1636.5 | 1.9231 | 4.9182 | 1454.0 | 1636.1 | 1.9113 | 4.3689 | 1453.6 | 1635.6 | 1.8981 |
| 1400 | 6.1377 | 1541.6 | 1746.0 | 1.9855 | 5.5222 | 1541.4 | 1745.7 | 1.9737 | 4.9068 | 1541.1 | 1745.4 | 1.9606 |
| 1600 | 6.8054 | 1632.4 | 1859.1 | 2.0432 | 6.1238 | 1632.2 | 1858.8 | 2.0315 | 5.4422 | 1632.0 | 1858.6 | 2.0184 |
| 1800 | 7.4716 | 1726.7 | 1975.6 | 2.0971 | 6.7238 | 1726.5 | 1975.4 | 2.0855 | 5.9760 | 1726.4 | 1975.2 | 2.0724 |
| 2000 | 8.1367 | 1824.4 | 2095.4 | 2.1479 | 7.3227 | 1824.3 | 2095.3 | 2.1363 | 6.5087 | 1824.1 | 2095.1 | 2.1232 |
| | P = 250 psia (400.97°F) | | | | P = 275 psia (409.45°F) | | | | P = 300 psia (417.35°F) | | | |
| Sat. | 1.8440 | 1116.3 | 1201.6 | 1.5270 | 1.6806 | 1117.0 | 1202.6 | 1.5187 | 1.5435 | 1117.7 | 1203.3 | 1.5111 |
| 450 | 2.0027 | 1141.3 | 1234.0 | 1.5636 | 1.8034 | 1138.5 | 1230.3 | 1.5499 | 1.6369 | 1135.6 | 1226.4 | 1.5369 |
| 500 | 2.1506 | 1164.1 | 1263.6 | 1.5953 | 1.9415 | 1162.0 | 1260.8 | 1.5825 | 1.7670 | 1159.8 | 1257.9 | 1.5706 |
| 550 | 2.2910 | 1185.6 | 1291.5 | 1.6237 | 2.0715 | 1183.9 | 1289.3 | 1.6115 | 1.8885 | 1182.1 | 1287.0 | 1.6001 |
| 600 | 2.4264 | 1206.3 | 1318.6 | 1.6499 | 2.1964 | 1204.9 | 1316.7 | 1.6380 | 2.0046 | 1203.5 | 1314.8 | 1.6270 |
| 650 | 2.5586 | 1226.8 | 1345.1 | 1.6743 | 2.3179 | 1225.6 | 1343.5 | 1.6627 | 2.1172 | 1224.4 | 1341.9 | 1.6520 |
| 700 | 2.6883 | 1247.0 | 1371.4 | 1.6974 | 2.4369 | 1246.0 | 1370.0 | 1.6860 | 2.2273 | 1244.9 | 1368.6 | 1.6755 |
| 800 | 2.9429 | 1287.3 | 1423.5 | 1.7406 | 2.6699 | 1286.5 | 1422.4 | 1.7294 | 2.4424 | 1285.7 | 1421.3 | 1.7192 |
| 900 | 3.1930 | 1327.9 | 1475.6 | 1.7804 | 2.8984 | 1327.3 | 1474.8 | 1.7694 | 2.6529 | 1326.6 | 1473.9 | 1.7593 |
| 1000 | 3.4403 | 1369.0 | 1528.2 | 1.8177 | 3.1241 | 1368.5 | 1527.4 | 1.8068 | 2.8605 | 1367.9 | 1526.7 | 1.7968 |
| 1200 | 3.9295 | 1453.3 | 1635.0 | 1.8863 | 3.5700 | 1452.9 | 1634.5 | 1.8755 | 3.2704 | 1452.5 | 1634.0 | 1.8657 |
| 1400 | 4.4144 | 1540.8 | 1745.0 | 1.9488 | 4.0116 | 1540.5 | 1744.6 | 1.9381 | 3.6759 | 1540.2 | 1744.2 | 1.9284 |
| 1600 | 4.8969 | 1631.7 | 1858.3 | 2.0066 | 4.4507 | 1631.5 | 1858.0 | 1.9960 | 4.0789 | 1631.3 | 1857.7 | 1.9863 |
| 1800 | 5.3777 | 1726.2 | 1974.9 | 2.0607 | 4.8882 | 1726.0 | 1974.7 | 2.0501 | 4.4803 | 1725.8 | 1974.5 | 2.0404 |
| 2000 | 5.8575 | 1823.9 | 2094.9 | 2.1116 | 5.3247 | 1823.8 | 2094.7 | 2.1010 | 4.8807 | 1823.6 | 2094.6 | 2.0913 |

Superheated water *(Continued)*

| T | υ | u | h | s | υ | u | h | s | υ | u | h | s |
|---|---|---|---|---|---|---|---|---|---|---|---|---|
| °F | ft³/lbm | Btu/lbm | Btu/lbm | Btu/lbm·R | ft³/lbm | Btu/lbm | Btu/lbm | Btu/lbm·R | ft³/lbm | Btu/lbm | Btu/lbm | Btu/lbm·R |
| | *P* = 350 psia (431.74°F) | | | | *P* = 400 psia (444.62°F) | | | | *P* = 450 psia (456.31°F) | | | |
| Sat. | 1.3263 | 1118.5 | 1204.4 | 1.4973 | 1.1617 | 1119.0 | 1205.0 | 1.4852 | 1.0324 | 1119.2 | 1205.2 | 1.4742 |
| 450 | 1.3739 | 1129.3 | 1218.3 | 1.5128 | 1.1747 | 1122.5 | 1209.4 | 1.4901 | | | | |
| 500 | 1.4921 | 1155.2 | 1251.9 | 1.5487 | 1.2851 | 1150.4 | 1245.6 | 1.5288 | 1.1233 | 1145.4 | 1238.9 | 1.5103 |
| 550 | 1.6004 | 1178.6 | 1282.2 | 1.5795 | 1.3840 | 1174.9 | 1277.3 | 1.5610 | 1.2152 | 1171.1 | 1272.3 | 1.5441 |
| 600 | 1.7030 | 1200.6 | 1310.9 | 1.6073 | 1.4765 | 1197.6 | 1306.9 | 1.5897 | 1.3001 | 1194.6 | 1302.8 | 1.5737 |
| 650 | 1.8018 | 1221.9 | 1338.6 | 1.6328 | 1.5650 | 1219.4 | 1335.3 | 1.6158 | 1.3807 | 1216.9 | 1331.9 | 1.6005 |
| 700 | 1.8979 | 1242.8 | 1365.8 | 1.6567 | 1.6507 | 1240.7 | 1362.9 | 1.6401 | 1.4584 | 1238.5 | 1360.0 | 1.6253 |
| 800 | 2.0848 | 1284.1 | 1419.1 | 1.7009 | 1.8166 | 1282.5 | 1417.0 | 1.6849 | 1.6080 | 1280.8 | 1414.7 | 1.6706 |
| 900 | 2.2671 | 1325.3 | 1472.2 | 1.7414 | 1.9777 | 1324.0 | 1470.4 | 1.7257 | 1.7526 | 1322.7 | 1468.6 | 1.7117 |
| 1000 | 2.4464 | 1366.9 | 1525.3 | 1.7791 | 2.1358 | 1365.8 | 1523.9 | 1.7636 | 1.8942 | 1364.7 | 1522.4 | 1.7499 |
| 1200 | 2.7996 | 1451.7 | 1633.0 | 1.8483 | 2.4465 | 1450.9 | 1632.0 | 1.8331 | 2.1718 | 1450.1 | 1631.0 | 1.8196 |
| 1400 | 3.1484 | 1539.6 | 1743.5 | 1.9111 | 2.7527 | 1539.0 | 1742.7 | 1.8960 | 2.4450 | 1538.4 | 1742.0 | 1.8827 |
| 1600 | 3.4947 | 1630.8 | 1857.1 | 1.9691 | 3.0565 | 1630.3 | 1856.5 | 1.9541 | 2.7157 | 1629.8 | 1856.0 | 1.9409 |
| 1800 | 3.8394 | 1725.4 | 1974.0 | 2.0233 | 3.3586 | 1725.0 | 1973.6 | 2.0084 | 2.9847 | 1724.6 | 1973.2 | 1.9952 |
| 2000 | 4.1830 | 1823.3 | 2094.2 | 2.0742 | 3.6597 | 1823.0 | 2093.9 | 2.0594 | 3.2527 | 1822.6 | 2093.5 | 2.0462 |
| | *P* = 500 psia (467.04°F) | | | | *P* = 600 psia (486.24°F) | | | | *P* = 700 psia (503.13°F) | | | |
| Sat. | 0.92815 | 1119.1 | 1205.0 | 1.4642 | 0.77020 | 1118.3 | 1203.9 | 1.4463 | 0.65589 | 1116.9 | 1201.9 | 1.4305 |
| 500 | 0.99304 | 1140.1 | 1231.9 | 1.4928 | 0.79526 | 1128.2 | 1216.5 | 1.4596 | | | | |
| 550 | 1.07974 | 1167.1 | 1267.0 | 1.5284 | 0.87542 | 1158.7 | 1255.9 | 1.4996 | 0.72799 | 1149.5 | 1243.8 | 1.4730 |
| 600 | 1.15876 | 1191.4 | 1298.6 | 1.5590 | 0.94605 | 1184.9 | 1289.9 | 1.5325 | 0.79332 | 1177.9 | 1280.7 | 1.5087 |
| 650 | 1.23312 | 1214.3 | 1328.4 | 1.5865 | 1.01133 | 1209.0 | 1321.3 | 1.5614 | 0.85242 | 1203.4 | 1313.8 | 1.5393 |
| 700 | 1.30440 | 1236.4 | 1357.0 | 1.6117 | 1.07316 | 1231.9 | 1351.0 | 1.5877 | 0.90769 | 1227.2 | 1344.8 | 1.5666 |
| 800 | 1.44097 | 1279.2 | 1412.5 | 1.6576 | 1.19038 | 1275.8 | 1408.0 | 1.6348 | 1.01125 | 1272.4 | 1403.4 | 1.6150 |
| 900 | 1.57252 | 1321.4 | 1466.9 | 1.6992 | 1.30230 | 1318.7 | 1463.3 | 1.6771 | 1.10921 | 1316.0 | 1459.7 | 1.6581 |
| 1000 | 1.70094 | 1363.6 | 1521.0 | 1.7376 | 1.41097 | 1361.4 | 1518.1 | 1.7160 | 1.20381 | 1359.2 | 1515.2 | 1.6974 |
| 1100 | 1.82726 | 1406.2 | 1575.3 | 1.7735 | 1.51749 | 1404.4 | 1572.9 | 1.7522 | 1.29621 | 1402.5 | 1570.4 | 1.7341 |
| 1200 | 1.95211 | 1449.4 | 1630.0 | 1.8075 | 1.62252 | 1447.8 | 1627.9 | 1.7865 | 1.38709 | 1446.2 | 1625.9 | 1.7685 |
| 1400 | 2.1988 | 1537.8 | 1741.2 | 1.8708 | 1.82957 | 1536.6 | 1739.7 | 1.8501 | 1.56580 | 1535.4 | 1738.2 | 1.8324 |
| 1600 | 2.4430 | 1629.4 | 1855.4 | 1.9291 | 2.0340 | 1628.4 | 1854.2 | 1.9085 | 1.74192 | 1627.5 | 1853.1 | 1.8911 |
| 1800 | 2.6856 | 1724.2 | 1972.7 | 1.9834 | 2.2369 | 1723.4 | 1971.8 | 1.9630 | 1.91643 | 1722.7 | 1970.9 | 1.9457 |
| 2000 | 2.9271 | 1822.3 | 2093.1 | 2.0345 | 2.4387 | 1821.7 | 2092.4 | 2.0141 | 2.08987 | 1821.0 | 2091.7 | 1.9969 |
| | *P* = 800 psia (518.27°F) | | | | *P* = 1000 psia (544.65°F) | | | | *P* = 1250 psia (572.45°F) | | | |
| Sat. | 0.56920 | 1115.0 | 1199.3 | 1.4162 | 0.44604 | 1110.1 | 1192.6 | 1.3906 | 0.34549 | 1102.0 | 1181.9 | 1.3623 |
| 550 | 0.61586 | 1139.4 | 1230.5 | 1.4476 | 0.45375 | 1115.2 | 1199.2 | 1.3972 | | | | |
| 600 | 0.67799 | 1170.5 | 1270.9 | 1.4866 | 0.51431 | 1154.1 | 1249.3 | 1.4457 | 0.37894 | 1129.5 | 1217.2 | 1.3961 |
| 650 | 0.73279 | 1197.6 | 1306.0 | 1.5191 | 0.56411 | 1185.1 | 1289.5 | 1.4827 | 0.42703 | 1167.5 | 1266.3 | 1.4414 |
| 700 | 0.78330 | 1222.4 | 1338.4 | 1.5476 | 0.60844 | 1212.4 | 1325.0 | 1.5140 | 0.46735 | 1198.7 | 1306.8 | 1.4771 |
| 750 | 0.83102 | 1246.0 | 1369.1 | 1.5735 | 0.64944 | 1237.6 | 1357.8 | 1.5418 | 0.50344 | 1226.4 | 1342.9 | 1.5076 |
| 800 | 0.87678 | 1268.9 | 1398.7 | 1.5975 | 0.68821 | 1261.7 | 1389.0 | 1.5670 | 0.53687 | 1252.2 | 1376.4 | 1.5347 |
| 900 | 0.96434 | 1313.3 | 1456.0 | 1.6413 | 0.76136 | 1307.7 | 1448.6 | 1.6126 | 0.59876 | 1300.5 | 1439.0 | 1.5826 |
| 1000 | 1.04841 | 1357.0 | 1512.2 | 1.6812 | 0.83078 | 1352.5 | 1506.2 | 1.6535 | 0.65656 | 1346.7 | 1498.6 | 1.6249 |
| 1100 | 1.13024 | 1400.7 | 1568.0 | 1.7181 | 0.89783 | 1396.9 | 1563.1 | 1.6911 | 0.71184 | 1392.2 | 1556.8 | 1.6635 |
| 1200 | 1.21051 | 1444.6 | 1623.8 | 1.7528 | 0.96327 | 1441.4 | 1619.7 | 1.7263 | 0.76545 | 1437.4 | 1614.5 | 1.6993 |
| 1400 | 1.36797 | 1534.2 | 1736.7 | 1.8170 | 1.09101 | 1531.8 | 1733.7 | 1.7911 | 0.86944 | 1528.7 | 1729.8 | 1.7649 |
| 1600 | 1.52283 | 1626.5 | 1851.9 | 1.8759 | 1.21610 | 1624.6 | 1849.6 | 1.8504 | 0.97072 | 1622.2 | 1846.7 | 1.8246 |
| 1800 | 1.67606 | 1721.9 | 1970.0 | 1.9306 | 1.33956 | 1720.3 | 1968.2 | 1.9053 | 1.07036 | 1718.4 | 1966.0 | 1.8799 |
| 2000 | 1.82823 | 1820.4 | 2091.0 | 1.9819 | 1.46194 | 1819.1 | 2089.6 | 1.9568 | 1.16892 | 1817.5 | 2087.9 | 1.9315 |

Superheated water *(Continued)*

| T °F | υ ft³/lbm | u Btu/lbm | h Btu/lbm | s Btu/lbm·R | υ ft³/lbm | u Btu/lbm | h Btu/lbm | s Btu/lbm·R | υ ft³/lbm | u Btu/lbm | h Btu/lbm | s Btu/lbm·R |
|---|---|---|---|---|---|---|---|---|---|---|---|---|
| | *P* = 1500 psia (596.26°F) | | | | *P* = 1750 psia (617.17°F) | | | | *P* = 2000 psia (635.85°F) | | | |
| Sat. | 0.27695 | 1092.1 | 1169.0 | 1.3362 | 0.22681 | 1080.5 | 1153.9 | 1.3112 | 0.18815 | 1066.8 | 1136.4 | 1.2863 |
| 600 | 0.28189 | 1097.2 | 1175.4 | 1.3423 | | | | | | | | |
| 650 | 0.33310 | 1147.2 | 1239.7 | 1.4016 | 0.26292 | 1122.8 | 1207.9 | 1.3607 | 0.20586 | 1091.4 | 1167.6 | 1.3146 |
| 700 | 0.37198 | 1183.6 | 1286.9 | 1.4433 | 0.30252 | 1166.8 | 1264.7 | 1.4108 | 0.24894 | 1147.6 | 1239.8 | 1.3783 |
| 750 | 0.40535 | 1214.4 | 1326.9 | 1.4771 | 0.33455 | 1201.5 | 1309.8 | 1.4489 | 0.28074 | 1187.4 | 1291.3 | 1.4218 |
| 800 | 0.43550 | 1242.2 | 1363.1 | 1.5064 | 0.36266 | 1231.7 | 1349.1 | 1.4807 | 0.30763 | 1220.5 | 1334.3 | 1.4567 |
| 850 | 0.46356 | 1268.2 | 1396.9 | 1.5328 | 0.38835 | 1259.3 | 1385.1 | 1.5088 | 0.33169 | 1250.0 | 1372.8 | 1.4867 |
| 900 | 0.49015 | 1293.1 | 1429.2 | 1.5569 | 0.41238 | 1285.4 | 1419.0 | 1.5341 | 0.35390 | 1277.5 | 1408.5 | 1.5134 |
| 1000 | 0.54031 | 1340.9 | 1490.8 | 1.6007 | 0.45719 | 1334.9 | 1482.9 | 1.5796 | 0.39479 | 1328.7 | 1474.9 | 1.5606 |
| 1100 | 0.58781 | 1387.3 | 1550.5 | 1.6402 | 0.49917 | 1382.4 | 1544.1 | 1.6201 | 0.43266 | 1377.5 | 1537.6 | 1.6021 |
| 1200 | 0.63355 | 1433.3 | 1609.2 | 1.6767 | 0.53932 | 1429.2 | 1603.9 | 1.6572 | 0.46864 | 1425.1 | 1598.5 | 1.6400 |
| 1400 | 0.72172 | 1525.7 | 1726.0 | 1.7432 | 0.61621 | 1522.6 | 1722.1 | 1.7245 | 0.53708 | 1519.5 | 1718.3 | 1.7081 |
| 1600 | 0.80714 | 1619.8 | 1843.8 | 1.8033 | 0.69031 | 1617.4 | 1840.9 | 1.7852 | 0.60269 | 1615.0 | 1838.0 | 1.7693 |
| 1800 | 0.89090 | 1716.4 | 1963.7 | 1.8589 | 0.76273 | 1714.5 | 1961.5 | 1.8410 | 0.66660 | 1712.5 | 1959.2 | 1.8255 |
| 2000 | 0.97358 | 1815.9 | 2086.1 | 1.9108 | 0.83406 | 1814.2 | 2084.3 | 1.8931 | 0.72942 | 1812.6 | 2082.6 | 1.8778 |
| | *P* = 2500 psia (668.17°F) | | | | *P* = 3000 psia (695.41°F) | | | | *P* = 3500 psia | | | |
| Sat. | 0.13076 | 1031.2 | 1091.7 | 1.2330 | 0.08460 | 969.81 | 1016.8 | 1.1587 | | | | |
| 650 | | | | | | | | | 0.02492 | 663.7 | 679.9 | 0.8632 |
| 700 | 0.16849 | 1098.4 | 1176.3 | 1.3072 | 0.09838 | 1005.3 | 1059.9 | 1.1960 | 0.03065 | 760.0 | 779.9 | 0.9511 |
| 750 | 0.20327 | 1154.9 | 1249.0 | 1.3686 | 0.14840 | 1114.1 | 1196.5 | 1.3118 | 0.10460 | 1057.6 | 1125.4 | 1.2434 |
| 800 | 0.22949 | 1195.9 | 1302.0 | 1.4116 | 0.17601 | 1167.5 | 1265.3 | 1.3676 | 0.13639 | 1134.3 | 1222.6 | 1.3224 |
| 850 | 0.25174 | 1230.1 | 1346.6 | 1.4463 | 0.19771 | 1208.2 | 1317.9 | 1.4086 | 0.15847 | 1183.8 | 1286.5 | 1.3721 |
| 900 | 0.27165 | 1260.7 | 1386.4 | 1.4761 | 0.21640 | 1242.8 | 1362.9 | 1.4423 | 0.17659 | 1223.4 | 1337.8 | 1.4106 |
| 950 | 0.29001 | 1289.1 | 1423.3 | 1.5028 | 0.23321 | 1273.9 | 1403.3 | 1.4716 | 0.19245 | 1257.8 | 1382.4 | 1.4428 |
| 1000 | 0.30726 | 1316.1 | 1458.2 | 1.5271 | 0.24876 | 1302.8 | 1440.9 | 1.4978 | 0.20687 | 1289.0 | 1423.0 | 1.4711 |
| 1100 | 0.33949 | 1367.3 | 1524.4 | 1.5710 | 0.27732 | 1356.8 | 1510.8 | 1.5441 | 0.23289 | 1346.1 | 1496.9 | 1.5201 |
| 1200 | 0.36966 | 1416.6 | 1587.6 | 1.6103 | 0.30367 | 1408.0 | 1576.6 | 1.5850 | 0.25654 | 1399.3 | 1565.4 | 1.5627 |
| 1400 | 0.42631 | 1513.3 | 1710.5 | 1.6802 | 0.35249 | 1507.0 | 1702.7 | 1.6567 | 0.29978 | 1500.7 | 1694.8 | 1.6364 |
| 1600 | 0.48004 | 1610.1 | 1832.2 | 1.7424 | 0.39830 | 1605.3 | 1826.4 | 1.7199 | 0.33994 | 1600.4 | 1820.5 | 1.7006 |
| 1800 | 0.53205 | 1708.6 | 1954.8 | 1.7991 | 0.44237 | 1704.7 | 1950.3 | 1.7773 | 0.37833 | 1700.8 | 1945.8 | 1.7586 |
| 2000 | 0.58295 | 1809.4 | 2079.1 | 1.8518 | 0.48532 | 1806.1 | 2075.6 | 1.8304 | 0.41561 | 1802.9 | 2072.1 | 1.8121 |
| | *P* = 4000 psia | | | | *P* = 5000 psia | | | | *P* = 6000 psia | | | |
| 650 | 0.02448 | 657.9 | 676.1 | 0.8577 | 0.02379 | 648.3 | 670.3 | 0.8485 | 0.02325 | 640.3 | 666.1 | 0.8408 |
| 700 | 0.02871 | 742.3 | 763.6 | 0.9347 | 0.02678 | 721.8 | 746.6 | 0.9156 | 0.02564 | 708.1 | 736.5 | 0.9028 |
| 750 | 0.06370 | 962.1 | 1009.2 | 1.1410 | 0.03373 | 821.8 | 853.0 | 1.0054 | 0.02981 | 788.7 | 821.8 | 0.9747 |
| 800 | 0.10520 | 1094.2 | 1172.1 | 1.2734 | 0.05937 | 986.9 | 1041.8 | 1.1581 | 0.03949 | 897.1 | 941.0 | 1.0711 |
| 850 | 0.12848 | 1156.7 | 1251.8 | 1.3355 | 0.08551 | 1092.4 | 1171.5 | 1.2593 | 0.05815 | 1018.6 | 1083.1 | 1.1819 |
| 900 | 0.14647 | 1202.5 | 1310.9 | 1.3799 | 0.10390 | 1155.9 | 1252.1 | 1.3198 | 0.07584 | 1103.5 | 1187.7 | 1.2603 |
| 950 | 0.16176 | 1240.7 | 1360.5 | 1.4157 | 0.11863 | 1203.9 | 1313.6 | 1.3643 | 0.09010 | 1163.7 | 1263.7 | 1.3153 |
| 1000 | 0.17538 | 1274.6 | 1404.4 | 1.4463 | 0.13128 | 1244.0 | 1365.5 | 1.4004 | 0.10208 | 1211.4 | 1324.7 | 1.3578 |
| 1100 | 0.19957 | 1335.1 | 1482.8 | 1.4983 | 0.15298 | 1312.2 | 1453.8 | 1.4590 | 0.12211 | 1288.4 | 1424.0 | 1.4237 |
| 1200 | 0.22121 | 1390.3 | 1554.1 | 1.5426 | 0.17185 | 1372.1 | 1531.1 | 1.5070 | 0.13911 | 1353.4 | 1507.8 | 1.4758 |
| 1300 | 0.24128 | 1443.0 | 1621.6 | 1.5821 | 0.18902 | 1427.8 | 1602.7 | 1.5490 | 0.15434 | 1412.5 | 1583.8 | 1.5203 |
| 1400 | 0.26028 | 1494.3 | 1687.0 | 1.6182 | 0.20508 | 1481.4 | 1671.1 | 1.5868 | 0.16841 | 1468.4 | 1655.4 | 1.5598 |
| 1600 | 0.29620 | 1595.5 | 1814.7 | 1.6835 | 0.23505 | 1585.6 | 1803.1 | 1.6542 | 0.19438 | 1575.7 | 1791.5 | 1.6294 |
| 1800 | 0.33033 | 1696.8 | 1941.4 | 1.7422 | 0.26320 | 1689.0 | 1932.5 | 1.7142 | 0.21853 | 1681.1 | 1923.7 | 1.6907 |
| 2000 | 0.36335 | 1799.7 | 2068.6 | 1.7961 | 0.29023 | 1793.2 | 2061.7 | 1.7689 | 0.24155 | 1786.7 | 2054.9 | 1.7463 |

## TABLE A–7E

Compressed liquid water

| T °F | υ ft³/lbm | u Btu/lbm | h Btu/lbm | s Btu/lbm·R | υ ft³/lbm | u Btu/lbm | h Btu/lbm | s Btu/lbm·R | υ ft³/lbm | u Btu/lbm | h Btu/lbm | s Btu/lbm·R |
|---|---|---|---|---|---|---|---|---|---|---|---|---|
| | *P* = 500 psia (467.04°F) | | | | *P* = 1000 psia (544.65°F) | | | | *P* = 1500 psia (596.26°F) | | | |
| Sat. | 0.019750 | 447.68 | 449.51 | 0.64900 | 0.021595 | 538.58 | 542.57 | 0.74341 | 0.023456 | 605.07 | 611.58 | 0.80836 |
| 32 | 0.015994 | 0.01 | 1.49 | 0.00001 | 0.015966 | 0.03 | 2.99 | 0.00005 | 0.015939 | 0.05 | 4.48 | 0.00008 |
| 50 | 0.015998 | 18.03 | 19.51 | 0.03601 | 0.015972 | 17.99 | 20.95 | 0.03593 | 0.015946 | 17.95 | 22.38 | 0.03584 |
| 100 | 0.016107 | 67.86 | 69.35 | 0.12930 | 0.016083 | 67.69 | 70.67 | 0.12899 | 0.016059 | 67.53 | 71.98 | 0.12869 |
| 150 | 0.016317 | 117.70 | 119.21 | 0.21462 | 0.016292 | 117.42 | 120.43 | 0.21416 | 0.016267 | 117.14 | 121.66 | 0.21369 |
| 200 | 0.016607 | 167.70 | 169.24 | 0.29349 | 0.016580 | 167.31 | 170.38 | 0.29289 | 0.016553 | 166.92 | 171.52 | 0.29229 |
| 250 | 0.016972 | 218.04 | 219.61 | 0.36708 | 0.016941 | 217.51 | 220.65 | 0.36634 | 0.016911 | 217.00 | 221.69 | 0.36560 |
| 300 | 0.017417 | 268.92 | 270.53 | 0.43641 | 0.017380 | 268.24 | 271.46 | 0.43551 | 0.017345 | 267.57 | 272.39 | 0.43463 |
| 350 | 0.017954 | 320.64 | 322.30 | 0.50240 | 0.017910 | 319.77 | 323.08 | 0.50132 | 0.017866 | 318.91 | 323.87 | 0.50025 |
| 400 | 0.018609 | 373.61 | 375.33 | 0.56595 | 0.018552 | 372.48 | 375.91 | 0.56463 | 0.018496 | 371.37 | 376.51 | 0.56333 |
| 450 | 0.019425 | 428.44 | 430.24 | 0.62802 | 0.019347 | 426.93 | 430.51 | 0.62635 | 0.019271 | 425.47 | 430.82 | 0.62472 |
| 500 | | | | | 0.020368 | 484.03 | 487.80 | 0.68764 | 0.020258 | 482.01 | 487.63 | 0.68550 |
| 550 | | | | | | | | | 0.021595 | 542.50 | 548.50 | 0.74731 |
| | *P* = 2000 psia (635.85°F) | | | | *P* = 3000 psia (695.41°F) | | | | *P* = 5000 psia | | | |
| Sat. | 0.025634 | 662.33 | 671.82 | 0.86224 | 0.034335 | 783.39 | 802.45 | 0.97321 | | | | |
| 32 | 0.015912 | 0.07 | 5.96 | 0.00010 | 0.015859 | 0.10 | 8.90 | 0.00011 | 0.015756 | 0.13 | 14.71 | 0.00002 |
| 50 | 0.015921 | 17.91 | 23.80 | 0.03574 | 0.015870 | 17.83 | 26.64 | 0.03554 | 0.015773 | 17.65 | 32.25 | 0.03505 |
| 100 | 0.016035 | 67.36 | 73.30 | 0.12838 | 0.015988 | 67.04 | 75.91 | 0.12776 | 0.015897 | 66.41 | 81.12 | 0.12652 |
| 200 | 0.016527 | 166.54 | 172.66 | 0.29170 | 0.016475 | 165.79 | 174.94 | 0.29053 | 0.016375 | 164.36 | 179.51 | 0.28824 |
| 300 | 0.017310 | 266.92 | 273.33 | 0.43376 | 0.017242 | 265.65 | 275.22 | 0.43204 | 0.017112 | 263.24 | 279.07 | 0.42874 |
| 400 | 0.018442 | 370.30 | 377.12 | 0.56205 | 0.018338 | 368.22 | 378.41 | 0.55959 | 0.018145 | 364.35 | 381.14 | 0.55492 |
| 450 | 0.019199 | 424.06 | 431.16 | 0.62314 | 0.019062 | 421.36 | 431.94 | 0.62010 | 0.018812 | 416.40 | 433.80 | 0.61445 |
| 500 | 0.020154 | 480.08 | 487.54 | 0.68346 | 0.019960 | 476.45 | 487.53 | 0.67958 | 0.019620 | 469.94 | 488.10 | 0.67254 |
| 560 | 0.021739 | 552.21 | 560.26 | 0.75692 | 0.021405 | 546.59 | 558.47 | 0.75126 | 0.020862 | 537.08 | 556.38 | 0.74154 |
| 600 | 0.023317 | 605.77 | 614.40 | 0.80898 | 0.022759 | 597.42 | 610.06 | 0.80086 | 0.021943 | 584.42 | 604.72 | 0.78803 |
| 640 | | | | | 0.024765 | 654.52 | 668.27 | 0.85476 | 0.023358 | 634.95 | 656.56 | 0.83603 |
| 680 | | | | | 0.028821 | 728.63 | 744.64 | 0.92288 | 0.025366 | 690.67 | 714.14 | 0.88745 |
| 700 | | | | | | | | | 0.026777 | 721.78 | 746.56 | 0.91564 |

Saturated ice–water vapor

| Temp., $T$ °F | Sat. press., $P_{sat}$ psia | Specific volume, ft³/lbm | | Internal energy, Btu/lbm | | | Enthalpy, Btu/lbm | | | Entropy, Btu/lbm·R | | |
|---|---|---|---|---|---|---|---|---|---|---|---|---|
| | | Sat. ice, $v_i$ | Sat. vapor, $v_g$ | Sat. ice, $u_i$ | Subl., $u_{ig}$ | Sat. vapor, $u_g$ | Sat. ice, $h_i$ | Subl., $h_{ig}$ | Sat. vapor, $h_g$ | Sat. ice, $s_i$ | Subl., $s_{ig}$ | Sat. vapor, $s_g$ |
| 32.018 | 0.08871 | 0.01747 | 3299.6 | −143.34 | 1164.2 | 1020.9 | −143.34 | 1218.3 | 1075.0 | −0.29146 | 2.4779 | 2.1864 |
| 32 | 0.08864 | 0.01747 | 3302.6 | −143.35 | 1164.2 | 1020.9 | −143.35 | 1218.4 | 1075.0 | −0.29148 | 2.4779 | 2.1865 |
| 30 | 0.08086 | 0.01747 | 3605.8 | −144.35 | 1164.6 | 1020.2 | −144.35 | 1218.5 | 1074.2 | −0.29353 | 2.4883 | 2.1948 |
| 25 | 0.06405 | 0.01746 | 4505.8 | −146.85 | 1165.4 | 1018.6 | −146.85 | 1218.8 | 1072.0 | −0.29865 | 2.5146 | 2.2160 |
| 20 | 0.05049 | 0.01746 | 5657.6 | −149.32 | 1166.2 | 1016.9 | −149.32 | 1219.1 | 1069.8 | −0.30377 | 2.5414 | 2.2376 |
| 15 | 0.03960 | 0.01745 | 7138.9 | −151.76 | 1167.0 | 1015.2 | −151.76 | 1219.3 | 1067.6 | −0.30889 | 2.5687 | 2.2598 |
| 10 | 0.03089 | 0.01744 | 9054.0 | −154.18 | 1167.8 | 1013.6 | −154.18 | 1219.5 | 1065.4 | −0.31401 | 2.5965 | 2.2825 |
| 5 | 0.02397 | 0.01743 | 11,543 | −156.57 | 1168.5 | 1011.9 | −156.57 | 1219.7 | 1063.1 | −0.31913 | 2.6248 | 2.3057 |
| 0 | 0.01850 | 0.01743 | 14,797 | −158.94 | 1169.2 | 1010.3 | −158.94 | 1219.9 | 1060.9 | −0.32426 | 2.6537 | 2.3295 |
| −5 | 0.01420 | 0.01742 | 19,075 | −161.28 | 1169.9 | 1008.6 | −161.28 | 1220.0 | 1058.7 | −0.32938 | 2.6832 | 2.3538 |
| −10 | 0.01083 | 0.01741 | 24,731 | −163.60 | 1170.6 | 1007.0 | −163.60 | 1220.1 | 1056.5 | −0.33451 | 2.7133 | 2.3788 |
| −15 | 0.00821 | 0.01740 | 32,257 | −165.90 | 1171.2 | 1005.3 | −165.90 | 1220.2 | 1054.3 | −0.33964 | 2.7440 | 2.4044 |
| −20 | 0.00619 | 0.01740 | 42,335 | −168.16 | 1171.8 | 1003.6 | −168.16 | 1220.3 | 1052.1 | −0.34478 | 2.7754 | 2.4306 |
| −25 | 0.00463 | 0.01739 | 55,917 | −170.41 | 1172.4 | 1002.0 | −170.41 | 1220.3 | 1049.9 | −0.34991 | 2.8074 | 2.4575 |
| −30 | 0.00344 | 0.01738 | 74,345 | −172.63 | 1173.0 | 1000.3 | −172.63 | 1220.3 | 1047.7 | −0.35505 | 2.8401 | 2.4850 |
| −35 | 0.00254 | 0.01738 | 99,526 | −174.83 | 1173.5 | 998.7 | −174.83 | 1220.3 | 1045.5 | −0.36019 | 2.8735 | 2.5133 |
| −40 | 0.00186 | 0.01737 | 134,182 | −177.00 | 1174.0 | 997.0 | −177.00 | 1220.3 | 1043.3 | −0.36534 | 2.9076 | 2.5423 |

**FIGURE A–9E**

*T-s* diagram for water.

*Source of Data: Joseph H. Keenan, Frederick G. Keyes, Philip G. Hill, and Joan G. Moore.* Steam Tables (*New York: John Wiley & Sons, 1969*)

**FIGURE A–10E**

Mollier diagram for water.

*Source of Data: Joseph H. Keenan, Frederick G. Keyes, Philip G. Hill, and Joan G. Moore.* Steam Tables *(New York: John Wiley & Sons, 1969)*

## TABLE A–11E

Saturated refrigerant-134a—Temperature table

| Temp., $T$ °F | Sat. press., $P_{sat}$ psia | Specific volume, ft³/lbm | | Internal energy, Btu/lbm | | | Enthalpy, Btu/lbm | | | Entropy, Btu/lbm·R | | |
|---|---|---|---|---|---|---|---|---|---|---|---|---|
| | | Sat. liquid, $v_f$ | Sat. vapor, $v_g$ | Sat. liquid, $u_f$ | Evap., $u_{fg}$ | Sat. vapor, $u_g$ | Sat. liquid, $h_f$ | Evap., $h_{fg}$ | Sat. vapor, $h_g$ | Sat. liquid, $s_f$ | Evap., $s_{fg}$ | Sat. vapor, $s_g$ |
| −40 | 7.432 | 0.01130 | 5.7769 | −0.016 | 89.174 | 89.16 | 0.000 | 97.104 | 97.10 | 0.00000 | 0.23136 | 0.23136 |
| −35 | 8.581 | 0.01136 | 5.0489 | 1.483 | 88.360 | 89.84 | 1.501 | 96.360 | 97.86 | 0.00355 | 0.22689 | 0.23044 |
| −30 | 9.869 | 0.01143 | 4.4286 | 2.987 | 87.542 | 90.53 | 3.008 | 95.608 | 98.62 | 0.00707 | 0.22250 | 0.22957 |
| −25 | 11.306 | 0.01149 | 3.8980 | 4.497 | 86.717 | 91.21 | 4.522 | 94.849 | 99.37 | 0.01057 | 0.21819 | 0.22876 |
| −20 | 12.906 | 0.01156 | 3.4424 | 6.014 | 85.887 | 91.90 | 6.041 | 94.080 | 100.12 | 0.01404 | 0.21396 | 0.22800 |
| −15 | 14.680 | 0.01163 | 3.0495 | 7.536 | 85.050 | 92.59 | 7.568 | 93.303 | 100.87 | 0.01748 | 0.20981 | 0.22729 |
| −10 | 16.642 | 0.01170 | 2.7097 | 9.065 | 84.206 | 93.27 | 9.102 | 92.515 | 101.62 | 0.02090 | 0.20572 | 0.22662 |
| −5 | 18.806 | 0.01178 | 2.4146 | 10.601 | 83.355 | 93.96 | 10.642 | 91.717 | 102.36 | 0.02430 | 0.20171 | 0.22600 |
| 0 | 21.185 | 0.01185 | 2.1575 | 12.143 | 82.496 | 94.64 | 12.190 | 90.907 | 103.10 | 0.02767 | 0.19775 | 0.22542 |
| 5 | 23.793 | 0.01193 | 1.9328 | 13.693 | 81.628 | 95.32 | 13.745 | 90.085 | 103.83 | 0.03103 | 0.19385 | 0.22488 |
| 10 | 26.646 | 0.01200 | 1.7358 | 15.249 | 80.751 | 96.00 | 15.308 | 89.251 | 104.56 | 0.03436 | 0.19001 | 0.22437 |
| 15 | 29.759 | 0.01208 | 1.5625 | 16.813 | 79.865 | 96.68 | 16.879 | 88.403 | 105.28 | 0.03767 | 0.18623 | 0.22390 |
| 20 | 33.147 | 0.01216 | 1.4097 | 18.384 | 78.969 | 97.35 | 18.459 | 87.541 | 106.00 | 0.04097 | 0.18249 | 0.22345 |
| 25 | 36.826 | 0.01225 | 1.2746 | 19.963 | 78.062 | 98.03 | 20.047 | 86.665 | 106.71 | 0.04424 | 0.17880 | 0.22304 |
| 30 | 40.813 | 0.01233 | 1.1548 | 21.550 | 77.144 | 98.69 | 21.643 | 85.772 | 107.42 | 0.04750 | 0.17515 | 0.22265 |
| 35 | 45.124 | 0.01242 | 1.0482 | 23.145 | 76.214 | 99.36 | 23.249 | 84.863 | 108.11 | 0.05074 | 0.17154 | 0.22228 |
| 40 | 49.776 | 0.01251 | 0.95323 | 24.749 | 75.272 | 100.02 | 24.864 | 83.937 | 108.80 | 0.05397 | 0.16797 | 0.22194 |
| 45 | 54.787 | 0.01261 | 0.86837 | 26.361 | 74.317 | 100.68 | 26.489 | 82.993 | 109.48 | 0.05718 | 0.16443 | 0.22162 |
| 50 | 60.175 | 0.01270 | 0.79236 | 27.983 | 73.347 | 101.33 | 28.124 | 82.029 | 110.15 | 0.06038 | 0.16093 | 0.22131 |
| 55 | 65.957 | 0.01280 | 0.72414 | 29.614 | 72.363 | 101.98 | 29.770 | 81.046 | 110.82 | 0.06357 | 0.15746 | 0.22103 |
| 60 | 72.152 | 0.01290 | 0.66277 | 31.254 | 71.364 | 102.62 | 31.426 | 80.041 | 111.47 | 0.06674 | 0.15401 | 0.22075 |
| 65 | 78.780 | 0.01301 | 0.60744 | 32.904 | 70.348 | 103.25 | 33.094 | 79.014 | 112.11 | 0.06991 | 0.15058 | 0.22049 |
| 70 | 85.858 | 0.01311 | 0.55746 | 34.565 | 69.315 | 103.88 | 34.773 | 77.964 | 112.74 | 0.07306 | 0.14718 | 0.22024 |
| 75 | 93.408 | 0.01323 | 0.51222 | 36.237 | 68.264 | 104.50 | 36.465 | 76.889 | 113.35 | 0.07621 | 0.14379 | 0.22000 |
| 80 | 101.45 | 0.01334 | 0.47119 | 37.920 | 67.193 | 105.11 | 38.170 | 75.788 | 113.96 | 0.07934 | 0.14042 | 0.21976 |
| 85 | 110.00 | 0.01346 | 0.43391 | 39.614 | 66.102 | 105.72 | 39.888 | 74.660 | 114.55 | 0.08247 | 0.13706 | 0.21953 |
| 90 | 119.08 | 0.01359 | 0.39997 | 41.321 | 64.989 | 106.31 | 41.620 | 73.503 | 115.12 | 0.08560 | 0.13371 | 0.21931 |
| 95 | 128.72 | 0.01372 | 0.36902 | 43.041 | 63.852 | 106.89 | 43.367 | 72.315 | 115.68 | 0.08872 | 0.13036 | 0.21908 |
| 100 | 138.93 | 0.01386 | 0.34074 | 44.774 | 62.690 | 107.46 | 45.130 | 71.094 | 116.22 | 0.09183 | 0.12702 | 0.21885 |
| 105 | 149.73 | 0.01400 | 0.31486 | 46.521 | 61.501 | 108.02 | 46.909 | 69.838 | 116.75 | 0.09495 | 0.12367 | 0.21862 |
| 110 | 161.16 | 0.01415 | 0.29113 | 48.284 | 60.284 | 108.57 | 48.706 | 68.544 | 117.25 | 0.09806 | 0.12031 | 0.21838 |
| 115 | 173.23 | 0.01430 | 0.26933 | 50.063 | 59.035 | 109.10 | 50.521 | 67.210 | 117.73 | 0.10118 | 0.11694 | 0.21813 |
| 120 | 185.96 | 0.01446 | 0.24928 | 51.858 | 57.753 | 109.61 | 52.356 | 65.833 | 118.19 | 0.10430 | 0.11356 | 0.21786 |
| 130 | 213.53 | 0.01482 | 0.21373 | 55.505 | 55.075 | 110.58 | 56.091 | 62.935 | 119.03 | 0.11056 | 0.10672 | 0.21728 |
| 140 | 244.06 | 0.01522 | 0.18331 | 59.237 | 52.221 | 111.46 | 59.925 | 59.813 | 119.74 | 0.11686 | 0.09973 | 0.21660 |
| 150 | 277.79 | 0.01567 | 0.15707 | 63.070 | 49.151 | 112.22 | 63.875 | 56.419 | 120.29 | 0.12324 | 0.09253 | 0.21577 |
| 160 | 314.94 | 0.01619 | 0.13423 | 67.022 | 45.811 | 112.83 | 67.965 | 52.690 | 120.66 | 0.12971 | 0.08502 | 0.21473 |
| 170 | 355.80 | 0.01682 | 0.11413 | 71.139 | 42.101 | 113.24 | 72.246 | 48.509 | 120.75 | 0.13637 | 0.07703 | 0.21340 |
| 180 | 400.66 | 0.01759 | 0.09619 | 75.464 | 37.893 | 113.36 | 76.768 | 43.721 | 120.49 | 0.14327 | 0.06834 | 0.21161 |
| 190 | 449.90 | 0.01861 | 0.07982 | 80.093 | 32.929 | 113.02 | 81.642 | 38.025 | 119.67 | 0.15057 | 0.05852 | 0.20909 |
| 200 | 504.00 | 0.02010 | 0.06441 | 85.297 | 26.629 | 111.93 | 87.172 | 30.761 | 117.93 | 0.15872 | 0.04662 | 0.20534 |
| 210 | 563.76 | 0.02309 | 0.04722 | 91.993 | 16.498 | 108.49 | 94.402 | 19.015 | 113.42 | 0.16924 | 0.02839 | 0.19763 |

*Source of Data:* Tables A-11E through A-13E are generated using the Engineering Equation Solver (EES) software developed by S. A. Klein and F. L. Alvarado. The routine used in calculations is the R134a, which is based on the fundamental equation of state developed by R. Tillner-Roth and H.D. Baehr, "An International Standard Formulation for the Thermodynamic Properties of 1,1,1,2-Tetrafluoroethane (HFC-134a) for temperatures from 170 K to 455 K and pressures up to 70 MPa," *J. Phys. Chem, Ref. Data,* Vol. 23, No. 5, 1994. The enthalpy and entropy values of saturated liquid are set to zero at −40°C (and −40°F).

## TABLE A-12E

Saturated refrigerant-134a—Pressure table

| Press., P psia | Sat. temp., $T_{sat}$ °F | Specific volume, ft³/lbm | | Internal energy, Btu/lbm | | | Enthalpy, Btu/lbm | | | Entropy, Btu/lbm·R | | |
|---|---|---|---|---|---|---|---|---|---|---|---|---|
| | | Sat. liquid, $v_f$ | Sat. vapor, $v_g$ | Sat. liquid, $u_f$ | Evap., $u_{fg}$ | Sat. vapor, $u_g$ | Sat. liquid, $h_f$ | Evap., $h_{fg}$ | Sat. vapor, $h_g$ | Sat. liquid, $s_f$ | Evap., $s_{fg}$ | Sat. vapor, $s_g$ |
| 5 | −53.09 | 0.01113 | 8.3740 | −3.914 | 91.283 | 87.37 | −3.903 | 99.021 | 95.12 | −0.00944 | 0.24353 | 0.23409 |
| 10 | −29.52 | 0.01143 | 4.3740 | 3.132 | 87.463 | 90.59 | 3.153 | 95.536 | 98.69 | 0.00741 | 0.22208 | 0.22949 |
| 15 | −14.15 | 0.01164 | 2.9882 | 7.796 | 84.907 | 92.70 | 7.828 | 93.170 | 101.00 | 0.01806 | 0.20911 | 0.22717 |
| 20 | −2.43 | 0.01181 | 2.2781 | 11.393 | 82.915 | 94.31 | 11.436 | 91.302 | 102.74 | 0.02603 | 0.19967 | 0.22570 |
| 25 | 7.17 | 0.01196 | 1.8442 | 14.367 | 81.249 | 95.62 | 14.422 | 89.725 | 104.15 | 0.03247 | 0.19218 | 0.22465 |
| 30 | 15.37 | 0.01209 | 1.5506 | 16.929 | 79.799 | 96.73 | 16.996 | 88.340 | 105.34 | 0.03792 | 0.18595 | 0.22386 |
| 35 | 22.57 | 0.01221 | 1.3382 | 19.195 | 78.504 | 97.70 | 19.274 | 87.093 | 106.37 | 0.04265 | 0.18058 | 0.22324 |
| 40 | 29.01 | 0.01232 | 1.1773 | 21.236 | 77.326 | 98.56 | 21.327 | 85.950 | 107.28 | 0.04686 | 0.17586 | 0.22272 |
| 45 | 34.86 | 0.01242 | 1.0510 | 23.101 | 76.240 | 99.34 | 23.205 | 84.889 | 108.09 | 0.05065 | 0.17164 | 0.22229 |
| 50 | 40.23 | 0.01252 | 0.94909 | 24.824 | 75.228 | 100.05 | 24.939 | 83.894 | 108.83 | 0.05412 | 0.16780 | 0.22192 |
| 55 | 45.20 | 0.01261 | 0.86509 | 26.428 | 74.277 | 100.70 | 26.556 | 82.954 | 109.51 | 0.05732 | 0.16429 | 0.22160 |
| 60 | 49.84 | 0.01270 | 0.79462 | 27.932 | 73.378 | 101.31 | 28.073 | 82.060 | 110.13 | 0.06028 | 0.16104 | 0.22132 |
| 65 | 54.20 | 0.01278 | 0.73462 | 29.351 | 72.523 | 101.87 | 29.505 | 81.205 | 110.71 | 0.06306 | 0.15801 | 0.22107 |
| 70 | 58.30 | 0.01287 | 0.68290 | 30.696 | 71.705 | 102.40 | 30.862 | 80.385 | 111.25 | 0.06567 | 0.15518 | 0.22084 |
| 75 | 62.19 | 0.01295 | 0.63784 | 31.975 | 70.921 | 102.90 | 32.155 | 79.594 | 111.75 | 0.06813 | 0.15251 | 0.22064 |
| 80 | 65.89 | 0.01303 | 0.59822 | 33.198 | 70.167 | 103.36 | 33.391 | 78.830 | 112.22 | 0.07047 | 0.14998 | 0.22045 |
| 85 | 69.41 | 0.01310 | 0.56309 | 34.369 | 69.438 | 103.81 | 34.575 | 78.089 | 112.66 | 0.07269 | 0.14758 | 0.22027 |
| 90 | 72.78 | 0.01318 | 0.53173 | 35.494 | 68.733 | 104.23 | 35.713 | 77.369 | 113.08 | 0.07481 | 0.14529 | 0.22011 |
| 95 | 76.02 | 0.01325 | 0.50356 | 36.577 | 68.048 | 104.63 | 36.810 | 76.668 | 113.48 | 0.07684 | 0.14311 | 0.21995 |
| 100 | 79.12 | 0.01332 | 0.47811 | 37.623 | 67.383 | 105.01 | 37.870 | 75.984 | 113.85 | 0.07879 | 0.14101 | 0.21981 |
| 110 | 85.00 | 0.01346 | 0.43390 | 39.614 | 66.102 | 105.72 | 39.888 | 74.660 | 114.55 | 0.08247 | 0.13706 | 0.21953 |
| 120 | 90.49 | 0.01360 | 0.39681 | 41.489 | 64.878 | 106.37 | 41.791 | 73.388 | 115.18 | 0.08590 | 0.13338 | 0.21928 |
| 130 | 95.64 | 0.01374 | 0.36523 | 43.263 | 63.704 | 106.97 | 43.594 | 72.159 | 115.75 | 0.08912 | 0.12993 | 0.21905 |
| 140 | 100.51 | 0.01387 | 0.33800 | 44.951 | 62.570 | 107.52 | 45.311 | 70.967 | 116.28 | 0.09215 | 0.12668 | 0.21883 |
| 150 | 105.12 | 0.01400 | 0.31426 | 46.563 | 61.473 | 108.04 | 46.952 | 69.807 | 116.76 | 0.09502 | 0.12359 | 0.21861 |
| 160 | 109.50 | 0.01413 | 0.29339 | 48.109 | 60.406 | 108.51 | 48.527 | 68.674 | 117.20 | 0.09776 | 0.12064 | 0.21840 |
| 170 | 113.69 | 0.01426 | 0.27487 | 49.595 | 59.366 | 108.96 | 50.043 | 67.564 | 117.61 | 0.10036 | 0.11783 | 0.21819 |
| 180 | 117.69 | 0.01439 | 0.25833 | 51.027 | 58.349 | 109.38 | 51.507 | 66.475 | 117.98 | 0.10286 | 0.11513 | 0.21799 |
| 190 | 121.53 | 0.01452 | 0.24346 | 52.412 | 57.353 | 109.76 | 52.922 | 65.402 | 118.32 | 0.10526 | 0.11252 | 0.21778 |
| 200 | 125.22 | 0.01464 | 0.23001 | 53.753 | 56.375 | 110.13 | 54.295 | 64.345 | 118.64 | 0.10757 | 0.11000 | 0.21757 |
| 220 | 132.21 | 0.01490 | 0.20662 | 56.321 | 54.462 | 110.78 | 56.927 | 62.267 | 119.19 | 0.11195 | 0.10519 | 0.21714 |
| 240 | 138.73 | 0.01516 | 0.18694 | 58.757 | 52.596 | 111.35 | 59.430 | 60.225 | 119.65 | 0.11606 | 0.10063 | 0.21669 |
| 260 | 144.85 | 0.01543 | 0.17012 | 61.082 | 50.763 | 111.84 | 61.824 | 58.205 | 120.03 | 0.11994 | 0.09627 | 0.21622 |
| 280 | 150.62 | 0.01570 | 0.15555 | 63.313 | 48.951 | 112.26 | 64.126 | 56.197 | 120.32 | 0.12364 | 0.09207 | 0.21571 |
| 300 | 156.09 | 0.01598 | 0.14279 | 65.460 | 47.154 | 112.61 | 66.347 | 54.195 | 120.54 | 0.12717 | 0.08800 | 0.21517 |
| 350 | 168.64 | 0.01672 | 0.11673 | 70.567 | 42.632 | 113.20 | 71.651 | 49.109 | 120.76 | 0.13545 | 0.07815 | 0.21360 |
| 400 | 179.86 | 0.01758 | 0.09643 | 75.401 | 37.957 | 113.36 | 76.702 | 43.794 | 120.50 | 0.14317 | 0.06847 | 0.21164 |
| 450 | 190.02 | 0.01860 | 0.07979 | 80.112 | 32.909 | 113.02 | 81.662 | 38.003 | 119.67 | 0.15060 | 0.05849 | 0.20909 |
| 500 | 199.29 | 0.01997 | 0.06533 | 84.900 | 27.096 | 112.00 | 86.748 | 31.292 | 118.04 | 0.15810 | 0.04748 | 0.20558 |

## TABLE A–13E

Superheated refrigerant-134a

| $T$ °F | $v$ ft³/lbm | $u$ Btu/lbm | $h$ Btu/lbm | $s$ Btu/lbm·R | $v$ ft³/lbm | $u$ Btu/lbm | $h$ Btu/lbm | $s$ Btu/lbm·R | $v$ ft³/lbm | $u$ Btu/lbm | $h$ Btu/lbm | $s$ Btu/lbm·R |
|---|---|---|---|---|---|---|---|---|---|---|---|---|
| | $P = 10$ psia ($T_{sat} = -29.52$°F) | | | | $P = 15$ psia ($T_{sat} = -14.15$°F) | | | | $P = 20$ psia ($T_{sat} = -2.43$°F) | | | |
| Sat. | 4.3740 | 90.59 | 98.69 | 0.22949 | 2.9882 | 92.70 | 101.00 | 0.22717 | 2.2781 | 94.31 | 102.74 | 0.22570 |
| −20 | 4.4856 | 92.14 | 100.44 | 0.23351 | | | | | | | | |
| 0 | 4.7135 | 95.42 | 104.14 | 0.24175 | 3.1001 | 95.08 | 103.69 | 0.23312 | 2.2922 | 94.73 | 103.21 | 0.22673 |
| 20 | 4.9380 | 98.77 | 107.91 | 0.24978 | 3.2551 | 98.49 | 107.52 | 0.24129 | 2.4130 | 98.19 | 107.12 | 0.23506 |
| 40 | 5.1600 | 102.21 | 111.76 | 0.25763 | 3.4074 | 101.96 | 111.42 | 0.24924 | 2.5306 | 101.71 | 111.07 | 0.24313 |
| 60 | 5.3802 | 105.73 | 115.68 | 0.26533 | 3.5577 | 105.51 | 115.38 | 0.25702 | 2.6461 | 105.29 | 115.08 | 0.25099 |
| 80 | 5.5989 | 109.33 | 119.69 | 0.27290 | 3.7064 | 109.14 | 119.42 | 0.26465 | 2.7600 | 108.94 | 119.15 | 0.25868 |
| 100 | 5.8165 | 113.02 | 123.78 | 0.28035 | 3.8540 | 112.85 | 123.54 | 0.27214 | 2.8726 | 112.67 | 123.30 | 0.26623 |
| 120 | 6.0331 | 116.80 | 127.96 | 0.28768 | 4.0006 | 116.64 | 127.75 | 0.27952 | 2.9842 | 116.48 | 127.53 | 0.27364 |
| 140 | 6.2490 | 120.66 | 132.23 | 0.29492 | 4.1464 | 120.52 | 132.03 | 0.28678 | 3.0950 | 120.38 | 131.83 | 0.28094 |
| 160 | 6.4642 | 124.62 | 136.58 | 0.30205 | 4.2915 | 124.49 | 136.40 | 0.29395 | 3.2051 | 124.35 | 136.22 | 0.28814 |
| 180 | 6.6789 | 128.66 | 141.01 | 0.30910 | 4.4361 | 128.53 | 140.85 | 0.30102 | 3.3146 | 128.41 | 140.68 | 0.29523 |
| 200 | 6.8930 | 132.78 | 145.54 | 0.31606 | 4.5802 | 132.67 | 145.38 | 0.30800 | 3.4237 | 132.56 | 145.23 | 0.30223 |
| 220 | 7.1068 | 136.99 | 150.14 | 0.32293 | 4.7239 | 136.89 | 150.00 | 0.31489 | 3.5324 | 136.78 | 149.86 | 0.30914 |
| | $P = 30$ psia ($T_{sat} = 15.37$°F) | | | | $P = 40$ psia ($T_{sat} = 29.01$°F) | | | | $P = 50$ psia ($T_{sat} = 40.23$°F) | | | |
| Sat. | 1.5506 | 96.73 | 105.34 | 0.22386 | 1.1773 | 98.56 | 107.28 | 0.22272 | 0.9491 | 100.05 | 108.83 | 0.22192 |
| 20 | 1.5691 | 97.56 | 106.27 | 0.22583 | | | | | | | | |
| 40 | 1.6528 | 101.18 | 110.35 | 0.23416 | 1.2126 | 100.61 | 109.59 | 0.22740 | | | | |
| 60 | 1.7338 | 104.83 | 114.45 | 0.24220 | 1.2768 | 104.35 | 113.80 | 0.23567 | 1.0019 | 103.85 | 113.12 | 0.23033 |
| 80 | 1.8130 | 108.54 | 118.60 | 0.25003 | 1.3389 | 108.12 | 118.03 | 0.24365 | 1.0540 | 107.69 | 117.44 | 0.23849 |
| 100 | 1.8908 | 112.31 | 122.81 | 0.25769 | 1.3995 | 111.94 | 122.30 | 0.25142 | 1.1043 | 111.56 | 121.78 | 0.24639 |
| 120 | 1.9675 | 116.16 | 127.08 | 0.26519 | 1.4588 | 115.83 | 126.63 | 0.25902 | 1.1534 | 115.49 | 126.16 | 0.25408 |
| 140 | 2.0434 | 120.08 | 131.43 | 0.27256 | 1.5173 | 119.79 | 131.02 | 0.26646 | 1.2015 | 119.48 | 130.60 | 0.26160 |
| 160 | 2.1185 | 124.09 | 135.85 | 0.27981 | 1.5750 | 123.82 | 135.47 | 0.27377 | 1.2488 | 123.54 | 135.09 | 0.26898 |
| 180 | 2.1931 | 128.17 | 140.34 | 0.28695 | 1.6321 | 127.92 | 140.00 | 0.28096 | 1.2955 | 127.67 | 139.66 | 0.27622 |
| 200 | 2.2671 | 132.33 | 144.92 | 0.29399 | 1.6887 | 132.10 | 144.60 | 0.28805 | 1.3416 | 131.87 | 144.29 | 0.28335 |
| 220 | 2.3408 | 136.58 | 149.57 | 0.30094 | 1.7449 | 136.37 | 149.28 | 0.29503 | 1.3873 | 136.15 | 148.99 | 0.29037 |
| 240 | 2.4141 | 140.90 | 154.30 | 0.30780 | 1.8007 | 140.70 | 154.03 | 0.30192 | 1.4326 | 140.51 | 153.76 | 0.29730 |
| 260 | 2.4871 | 145.30 | 159.11 | 0.31458 | 1.8562 | 145.12 | 158.86 | 0.30873 | 1.4776 | 144.94 | 158.61 | 0.30413 |
| 280 | 2.5598 | 149.79 | 164.00 | 0.32128 | 1.9114 | 149.62 | 163.77 | 0.31545 | 1.5223 | 149.45 | 163.53 | 0.31087 |
| | $P = 60$ psia ($T_{sat} = 49.84$°F) | | | | $P = 70$ psia ($T_{sat} = 58.30$°F) | | | | $P = 80$ psia ($T_{sat} = 65.89$°F) | | | |
| Sat. | 0.7946 | 101.31 | 110.13 | 0.22132 | 0.6829 | 102.40 | 111.25 | 0.22084 | 0.5982 | 103.36 | 112.22 | 0.22045 |
| 60 | 0.8179 | 103.31 | 112.39 | 0.22572 | 0.6857 | 102.74 | 111.62 | 0.22157 | | | | |
| 80 | 0.8636 | 107.24 | 116.82 | 0.23408 | 0.7271 | 106.77 | 116.18 | 0.23018 | 0.6243 | 106.27 | 115.51 | 0.22663 |
| 100 | 0.9072 | 111.17 | 121.24 | 0.24212 | 0.7662 | 110.77 | 120.69 | 0.23838 | 0.6601 | 110.35 | 120.12 | 0.23501 |
| 120 | 0.9495 | 115.14 | 125.69 | 0.24992 | 0.8037 | 114.79 | 125.20 | 0.24630 | 0.6941 | 114.43 | 124.70 | 0.24305 |
| 140 | 0.9908 | 119.17 | 130.17 | 0.25753 | 0.8401 | 118.86 | 129.74 | 0.25399 | 0.7270 | 118.53 | 129.29 | 0.25084 |
| 160 | 1.0312 | 123.26 | 134.71 | 0.26497 | 0.8756 | 122.98 | 134.32 | 0.26151 | 0.7589 | 122.69 | 133.92 | 0.25843 |
| 180 | 1.0709 | 127.42 | 139.31 | 0.27227 | 0.9105 | 127.16 | 138.95 | 0.26886 | 0.7900 | 126.89 | 138.59 | 0.26585 |
| 200 | 1.1101 | 131.64 | 143.97 | 0.27945 | 0.9447 | 131.40 | 143.64 | 0.27608 | 0.8206 | 131.17 | 143.31 | 0.27312 |
| 220 | 1.1489 | 135.94 | 148.69 | 0.28651 | 0.9785 | 135.72 | 148.40 | 0.28318 | 0.8507 | 135.50 | 148.09 | 0.28026 |
| 240 | 1.1872 | 140.31 | 153.49 | 0.29346 | 1.0118 | 140.11 | 153.22 | 0.29017 | 0.8803 | 139.91 | 152.94 | 0.28728 |
| 260 | 1.2252 | 144.76 | 158.36 | 0.30032 | 1.0449 | 144.57 | 158.10 | 0.29706 | 0.9096 | 144.38 | 157.85 | 0.29420 |
| 280 | 1.2629 | 149.28 | 163.30 | 0.30709 | 1.0776 | 149.10 | 163.06 | 0.30386 | 0.9386 | 148.93 | 162.82 | 0.30102 |
| 300 | 1.3004 | 153.88 | 168.31 | 0.31378 | 1.1101 | 153.71 | 168.09 | 0.31057 | 0.9674 | 153.55 | 167.87 | 0.30775 |
| 320 | 1.3377 | 158.55 | 173.40 | 0.32039 | 1.1424 | 158.40 | 173.20 | 0.31720 | 0.9959 | 158.25 | 172.99 | 0.31440 |

## TABLE A–13E

Superheated refrigerant-134a (Concluded)

| $T$ °F | $v$ ft³/lbm | $u$ Btu/lbm | $h$ Btu/lbm | $s$ Btu/lbm·R | $v$ ft³/lbm | $u$ Btu/lbm | $h$ Btu/lbm | $s$ Btu/lbm·R | $v$ ft³/lbm | $u$ Btu/lbm | $h$ Btu/lbm | $s$ Btu/lbm·R |
|---|---|---|---|---|---|---|---|---|---|---|---|---|
| | $P = 90$ psia ($T_{sat} = 72.78$°F) | | | | $P = 100$ psia ($T_{sat} = 79.12$°F) | | | | $P = 120$ psia ($T_{sat} = 90.49$°F) | | | |
| Sat. | 0.53173 | 104.23 | 113.08 | 0.22011 | 0.47811 | 105.01 | 113.85 | 0.21981 | 0.39681 | 106.37 | 115.18 | 0.21928 |
| 80 | 0.54388 | 105.75 | 114.81 | 0.22332 | 0.47906 | 105.19 | 114.06 | 0.22018 | | | | |
| 100 | 0.57729 | 109.91 | 119.53 | 0.23191 | 0.51076 | 109.46 | 118.91 | 0.22902 | 0.41013 | 108.49 | 117.59 | 0.22364 |
| 120 | 0.60874 | 114.05 | 124.19 | 0.24009 | 0.54022 | 113.66 | 123.66 | 0.23735 | 0.43692 | 112.85 | 122.55 | 0.23234 |
| 140 | 0.63885 | 118.20 | 128.84 | 0.24799 | 0.56821 | 117.86 | 128.38 | 0.24535 | 0.46190 | 117.16 | 127.42 | 0.24059 |
| 160 | 0.66796 | 122.39 | 133.51 | 0.25565 | 0.59513 | 122.09 | 133.10 | 0.25310 | 0.48563 | 121.47 | 132.25 | 0.24853 |
| 180 | 0.69629 | 126.63 | 138.22 | 0.26313 | 0.62122 | 126.36 | 137.85 | 0.26065 | 0.50844 | 125.80 | 137.09 | 0.25621 |
| 200 | 0.72399 | 130.92 | 142.98 | 0.27045 | 0.64667 | 130.68 | 142.64 | 0.26802 | 0.53054 | 130.18 | 141.96 | 0.26370 |
| 220 | 0.75119 | 135.28 | 147.79 | 0.27763 | 0.67158 | 135.05 | 147.48 | 0.27525 | 0.55206 | 134.60 | 146.86 | 0.27102 |
| 240 | 0.77796 | 139.70 | 152.66 | 0.28469 | 0.69605 | 139.50 | 152.38 | 0.28234 | 0.57312 | 139.08 | 151.80 | 0.27819 |
| 260 | 0.80437 | 144.19 | 157.59 | 0.29164 | 0.72016 | 144.00 | 157.33 | 0.28932 | 0.59379 | 143.62 | 156.80 | 0.28523 |
| 280 | 0.83048 | 148.75 | 162.58 | 0.29849 | 0.74396 | 148.58 | 162.34 | 0.29620 | 0.61413 | 148.22 | 161.86 | 0.29216 |
| 300 | 0.85633 | 153.39 | 167.65 | 0.30524 | 0.76749 | 153.22 | 167.42 | 0.30297 | 0.63420 | 152.89 | 166.97 | 0.29898 |
| 320 | 0.88195 | 158.09 | 172.78 | 0.31191 | 0.79079 | 157.94 | 172.57 | 0.30966 | 0.65402 | 157.62 | 172.15 | 0.30571 |
| | $P = 140$ psia ($T_{sat} = 100.51$°F) | | | | $P = 160$ psia ($T_{sat} = 109.50$°F) | | | | $P = 180$ psia ($T_{sat} = 117.69$°F) | | | |
| Sat. | 0.33800 | 107.52 | 116.28 | 0.21883 | 0.29339 | 108.51 | 117.20 | 0.21840 | 0.25833 | 109.38 | 117.98 | 0.21799 |
| 120 | 0.36243 | 111.97 | 121.36 | 0.22775 | 0.30578 | 111.01 | 120.07 | 0.22339 | 0.26083 | 109.95 | 118.64 | 0.21912 |
| 140 | 0.38551 | 116.42 | 126.40 | 0.23630 | 0.32774 | 115.63 | 125.33 | 0.23232 | 0.28231 | 114.78 | 124.18 | 0.22852 |
| 160 | 0.40711 | 120.82 | 131.37 | 0.24444 | 0.34790 | 120.14 | 130.44 | 0.24070 | 0.30154 | 119.43 | 129.47 | 0.23720 |
| 180 | 0.42766 | 125.23 | 136.31 | 0.25229 | 0.36686 | 124.63 | 135.49 | 0.24872 | 0.31936 | 124.01 | 134.65 | 0.24542 |
| 200 | 0.44743 | 129.66 | 141.25 | 0.25990 | 0.38494 | 129.13 | 140.52 | 0.25647 | 0.33619 | 128.58 | 139.77 | 0.25332 |
| 220 | 0.46657 | 134.13 | 146.22 | 0.26731 | 0.40234 | 133.65 | 145.56 | 0.26399 | 0.35228 | 133.16 | 144.89 | 0.26095 |
| 240 | 0.48522 | 138.65 | 151.22 | 0.27457 | 0.41921 | 138.21 | 150.62 | 0.27133 | 0.36779 | 137.76 | 150.01 | 0.26838 |
| 260 | 0.50345 | 143.22 | 156.26 | 0.28168 | 0.43564 | 142.82 | 155.72 | 0.27851 | 0.38284 | 142.41 | 155.16 | 0.27564 |
| 280 | 0.52134 | 147.85 | 161.36 | 0.28866 | 0.45171 | 147.48 | 160.86 | 0.28555 | 0.39751 | 147.11 | 160.35 | 0.28275 |
| 300 | 0.53895 | 152.55 | 166.51 | 0.29553 | 0.46748 | 152.21 | 166.05 | 0.29248 | 0.41186 | 151.86 | 165.58 | 0.28972 |
| 320 | 0.55630 | 157.31 | 171.72 | 0.30230 | 0.48299 | 156.99 | 171.29 | 0.29929 | 0.42594 | 156.67 | 170.85 | 0.29658 |
| 340 | 0.57345 | 162.14 | 176.99 | 0.30898 | 0.49828 | 161.84 | 176.59 | 0.30600 | 0.43980 | 161.53 | 176.18 | 0.30333 |
| 360 | 0.59041 | 167.03 | 182.33 | 0.31557 | 0.51338 | 166.75 | 181.95 | 0.31262 | 0.45347 | 166.47 | 181.57 | 0.30998 |
| | $P = 200$ psia ($T_{sat} = 125.22$°F) | | | | $P = 300$ psia ($T_{sat} = 156.09$°F) | | | | $P = 400$ psia ($T_{sat} = 179.86$°F) | | | |
| Sat. | 0.23001 | 110.13 | 118.64 | 0.21757 | 0.14279 | 112.61 | 120.54 | 0.21517 | 0.09643 | 113.36 | 120.50 | 0.21164 |
| 140 | 0.24541 | 113.86 | 122.94 | 0.22483 | | | | | | | | |
| 160 | 0.26412 | 118.67 | 128.44 | 0.23386 | 0.14656 | 113.82 | 121.96 | 0.21747 | | | | |
| 180 | 0.28115 | 123.36 | 133.77 | 0.24231 | 0.16355 | 119.53 | 128.61 | 0.22803 | 0.09658 | 113.42 | 120.56 | 0.21174 |
| 200 | 0.29704 | 128.01 | 139.00 | 0.25037 | 0.17776 | 124.79 | 134.66 | 0.23734 | 0.11440 | 120.53 | 128.99 | 0.22473 |
| 220 | 0.31212 | 132.65 | 144.20 | 0.25813 | 0.19044 | 129.86 | 140.43 | 0.24596 | 0.12746 | 126.45 | 135.88 | 0.23502 |
| 240 | 0.32658 | 137.31 | 149.39 | 0.26566 | 0.20211 | 134.83 | 146.05 | 0.25412 | 0.13853 | 131.96 | 142.21 | 0.24420 |
| 260 | 0.34054 | 141.99 | 154.60 | 0.27300 | 0.21306 | 139.77 | 151.60 | 0.26193 | 0.14844 | 137.27 | 148.26 | 0.25272 |
| 280 | 0.35410 | 146.73 | 159.83 | 0.28017 | 0.22347 | 144.71 | 157.11 | 0.26949 | 0.15756 | 142.48 | 154.15 | 0.26079 |
| 300 | 0.36733 | 151.50 | 165.10 | 0.28720 | 0.23346 | 149.66 | 162.62 | 0.27683 | 0.16611 | 147.65 | 159.95 | 0.26853 |
| 320 | 0.38029 | 156.34 | 170.41 | 0.29410 | 0.24310 | 154.63 | 168.13 | 0.28399 | 0.17423 | 152.81 | 165.71 | 0.27601 |
| 340 | 0.39300 | 161.23 | 175.77 | 0.30089 | 0.25246 | 159.65 | 173.66 | 0.29100 | 0.18201 | 157.97 | 171.45 | 0.28328 |
| 360 | 0.40552 | 166.18 | 181.19 | 0.30758 | 0.26159 | 164.71 | 179.23 | 0.29788 | 0.18951 | 163.16 | 177.19 | 0.29037 |

**FIGURE A–14E**

*P-h* diagram for refrigerant-134a.

## TABLE A–16E

Properties of the atmosphere at high altitude

| Altitude, ft | Temperature, °F | Pressure, psia | Gravity, g, ft/s² | Speed of sound, ft/s | Density, lbm/ft³ | Viscosity μ, lbm/ft·s | Thermal conductivity, Btu/h·ft·R |
|---|---|---|---|---|---|---|---|
| 0 | 59.00 | 14.7 | 32.174 | 1116 | 0.07647 | $1.202 \times 10^{-5}$ | 0.0146 |
| 500 | 57.22 | 14.4 | 32.173 | 1115 | 0.07536 | $1.199 \times 10^{-5}$ | 0.0146 |
| 1000 | 55.43 | 14.2 | 32.171 | 1113 | 0.07426 | $1.196 \times 10^{-5}$ | 0.0146 |
| 1500 | 53.65 | 13.9 | 32.169 | 1111 | 0.07317 | $1.193 \times 10^{-5}$ | 0.0145 |
| 2000 | 51.87 | 13.7 | 32.168 | 1109 | 0.07210 | $1.190 \times 10^{-5}$ | 0.0145 |
| 2500 | 50.09 | 13.4 | 32.166 | 1107 | 0.07104 | $1.186 \times 10^{-5}$ | 0.0144 |
| 3000 | 48.30 | 13.2 | 32.165 | 1105 | 0.06998 | $1.183 \times 10^{-5}$ | 0.0144 |
| 3500 | 46.52 | 12.9 | 32.163 | 1103 | 0.06985 | $1.180 \times 10^{-5}$ | 0.0143 |
| 4000 | 44.74 | 12.7 | 32.162 | 1101 | 0.06792 | $1.177 \times 10^{-5}$ | 0.0143 |
| 4500 | 42.96 | 12.5 | 32.160 | 1099 | 0.06690 | $1.173 \times 10^{-5}$ | 0.0142 |
| 5000 | 41.17 | 12.2 | 32.159 | 1097 | 0.06590 | $1.170 \times 10^{-5}$ | 0.0142 |
| 5500 | 39.39 | 12.0 | 32.157 | 1095 | 0.06491 | $1.167 \times 10^{-5}$ | 0.0141 |
| 6000 | 37.61 | 11.8 | 32.156 | 1093 | 0.06393 | $1.164 \times 10^{-5}$ | 0.0141 |
| 6500 | 35.83 | 11.6 | 32.154 | 1091 | 0.06296 | $1.160 \times 10^{-5}$ | 0.0141 |
| 7000 | 34.05 | 11.3 | 32.152 | 1089 | 0.06200 | $1.157 \times 10^{-5}$ | 0.0140 |
| 7500 | 32.26 | 11.1 | 32.151 | 1087 | 0.06105 | $1.154 \times 10^{-5}$ | 0.0140 |
| 8000 | 30.48 | 10.9 | 32.149 | 1085 | 0.06012 | $1.150 \times 10^{-5}$ | 0.0139 |
| 8500 | 28.70 | 10.7 | 32.148 | 1083 | 0.05919 | $1.147 \times 10^{-5}$ | 0.0139 |
| 9000 | 26.92 | 10.5 | 32.146 | 1081 | 0.05828 | $1.144 \times 10^{-5}$ | 0.0138 |
| 9500 | 25.14 | 10.3 | 32.145 | 1079 | 0.05738 | $1.140 \times 10^{-5}$ | 0.0138 |
| 10,000 | 23.36 | 10.1 | 32.145 | 1077 | 0.05648 | $1.137 \times 10^{-5}$ | 0.0137 |
| 11,000 | 19.79 | 9.72 | 32.140 | 1073 | 0.05473 | $1.130 \times 10^{-5}$ | 0.0136 |
| 12,000 | 16.23 | 9.34 | 32.137 | 1069 | 0.05302 | $1.124 \times 10^{-5}$ | 0.0136 |
| 13,000 | 12.67 | 8.99 | 32.134 | 1065 | 0.05135 | $1.117 \times 10^{-5}$ | 0.0135 |
| 14,000 | 9.12 | 8.63 | 32.131 | 1061 | 0.04973 | $1.110 \times 10^{-5}$ | 0.0134 |
| 15,000 | 5.55 | 8.29 | 32.128 | 1057 | 0.04814 | $1.104 \times 10^{-5}$ | 0.0133 |
| 16,000 | +1.99 | 7.97 | 32.125 | 1053 | 0.04659 | $1.097 \times 10^{-5}$ | 0.0132 |
| 17,000 | −1.58 | 7.65 | 32.122 | 1049 | 0.04508 | $1.090 \times 10^{-5}$ | 0.0132 |
| 18,000 | −5.14 | 7.34 | 32.119 | 1045 | 0.04361 | $1.083 \times 10^{-5}$ | 0.0130 |
| 19,000 | −8.70 | 7.05 | 32.115 | 1041 | 0.04217 | $1.076 \times 10^{-5}$ | 0.0129 |
| 20,000 | −12.2 | 6.76 | 32.112 | 1037 | 0.04077 | $1.070 \times 10^{-5}$ | 0.0128 |
| 22,000 | −19.4 | 6.21 | 32.106 | 1029 | 0.03808 | $1.056 \times 10^{-5}$ | 0.0126 |
| 24,000 | −26.5 | 5.70 | 32.100 | 1020 | 0.03553 | $1.042 \times 10^{-5}$ | 0.0124 |
| 26,000 | −33.6 | 5.22 | 32.094 | 1012 | 0.03311 | $1.028 \times 10^{-5}$ | 0.0122 |
| 28,000 | −40.7 | 4.78 | 32.088 | 1003 | 0.03082 | $1.014 \times 10^{-5}$ | 0.0121 |
| 30,000 | −47.8 | 4.37 | 32.082 | 995 | 0.02866 | $1.000 \times 10^{-5}$ | 0.0119 |
| 32,000 | −54.9 | 3.99 | 32.08 | 987 | 0.02661 | $0.986 \times 10^{-5}$ | 0.0117 |
| 34,000 | −62.0 | 3.63 | 32.07 | 978 | 0.02468 | $0.971 \times 10^{-5}$ | 0.0115 |
| 36,000 | −69.2 | 3.30 | 32.06 | 969 | 0.02285 | $0.956 \times 10^{-5}$ | 0.0113 |
| 38,000 | −69.7 | 3.05 | 32.06 | 968 | 0.02079 | $0.955 \times 10^{-5}$ | 0.0113 |
| 40,000 | −69.7 | 2.73 | 32.05 | 968 | 0.01890 | $0.955 \times 10^{-5}$ | 0.0113 |
| 45,000 | −69.7 | 2.148 | 32.04 | 968 | 0.01487 | $0.955 \times 10^{-5}$ | 0.0113 |
| 50,000 | −69.7 | 1.691 | 32.02 | 968 | 0.01171 | $0.955 \times 10^{-5}$ | 0.0113 |
| 55,000 | −69.7 | 1.332 | 32.00 | 968 | 0.00922 | $0.955 \times 10^{-5}$ | 0.0113 |
| 60,000 | −69.7 | 1.048 | 31.99 | 968 | 0.00726 | $0.955 \times 10^{-5}$ | 0.0113 |

*Source of Data:* U.S. Standard Atmosphere Supplements, U.S. Government Printing Office, 1966. Based on year-round mean conditions at 45° latitude and varies with the time of the year and the weather patterns. The conditions at sea level ($z = 0$) are taken to be $P = 14.696$ psia, $T = 59$°F, $\rho = 0.076474$ lbm/ft³, $g = 32.1741$ ft²/s.

Ideal-gas properties of air

| T R | h Btu/lbm | $P_r$ | u Btu/lbm | $v_r$ | $s°$ Btu/lbm·R | T R | h Btu/lbm | $P_r$ | u Btu/lbm | $v_r$ | $s°$ Btu/lbm·R |
|---|---|---|---|---|---|---|---|---|---|---|---|
| 360 | 85.97 | 0.3363 | 61.29 | 396.6 | 0.50369 | 1600 | 395.74 | 71.13 | 286.06 | 8.263 | 0.87130 |
| 380 | 90.75 | 0.4061 | 64.70 | 346.6 | 0.51663 | 1650 | 409.13 | 80.89 | 296.03 | 7.556 | 0.87954 |
| 400 | 95.53 | 0.4858 | 68.11 | 305.0 | 0.52890 | 1700 | 422.59 | 90.95 | 306.06 | 6.924 | 0.88758 |
| 420 | 100.32 | 0.5760 | 71.52 | 270.1 | 0.54058 | 1750 | 436.12 | 101.98 | 316.16 | 6.357 | 0.89542 |
| 440 | 105.11 | 0.6776 | 74.93 | 240.6 | 0.55172 | 1800 | 449.71 | 114.0 | 326.32 | 5.847 | 0.90308 |
| 460 | 109.90 | 0.7913 | 78.36 | 215.33 | 0.56235 | 1850 | 463.37 | 127.2 | 336.55 | 5.388 | 0.91056 |
| 480 | 114.69 | 0.9182 | 81.77 | 193.65 | 0.57255 | 1900 | 477.09 | 141.5 | 346.85 | 4.974 | 0.91788 |
| 500 | 119.48 | 1.0590 | 85.20 | 174.90 | 0.58233 | 1950 | 490.88 | 157.1 | 357.20 | 4.598 | 0.92504 |
| 520 | 124.27 | 1.2147 | 88.62 | 158.58 | 0.59173 | 2000 | 504.71 | 174.0 | 367.61 | 4.258 | 0.93205 |
| 537 | 128.10 | 1.3593 | 91.53 | 146.34 | 0.59945 | 2050 | 518.71 | 192.3 | 378.08 | 3.949 | 0.93891 |
| 540 | 129.06 | 1.3860 | 92.04 | 144.32 | 0.60078 | 2100 | 532.55 | 212.1 | 388.60 | 3.667 | 0.94564 |
| 560 | 133.86 | 1.5742 | 95.47 | 131.78 | 0.60950 | 2150 | 546.54 | 223.5 | 399.17 | 3.410 | 0.95222 |
| 580 | 138.66 | 1.7800 | 98.90 | 120.70 | 0.61793 | 2200 | 560.59 | 256.6 | 409.78 | 3.176 | 0.95919 |
| 600 | 143.47 | 2.005 | 102.34 | 110.88 | 0.62607 | 2250 | 574.69 | 281.4 | 420.46 | 2.961 | 0.96501 |
| 620 | 148.28 | 2.249 | 105.78 | 102.12 | 0.63395 | 2300 | 588.82 | 308.1 | 431.16 | 2.765 | 0.97123 |
| 640 | 153.09 | 2.514 | 109.21 | 94.30 | 0.64159 | 2350 | 603.00 | 336.8 | 441.91 | 2.585 | 0.97732 |
| 660 | 157.92 | 2.801 | 112.67 | 87.27 | 0.64902 | 2400 | 617.22 | 367.6 | 452.70 | 2.419 | 0.98331 |
| 680 | 162.73 | 3.111 | 116.12 | 80.96 | 0.65621 | 2450 | 631.48 | 400.5 | 463.54 | 2.266 | 0.98919 |
| 700 | 167.56 | 3.446 | 119.58 | 75.25 | 0.66321 | 2500 | 645.78 | 435.7 | 474.40 | 2.125 | 0.99497 |
| 720 | 172.39 | 3.806 | 123.04 | 70.07 | 0.67002 | 2550 | 660.12 | 473.3 | 485.31 | 1.996 | 1.00064 |
| 740 | 177.23 | 4.193 | 126.51 | 65.38 | 0.67665 | 2600 | 674.49 | 513.5 | 496.26 | 1.876 | 1.00623 |
| 760 | 182.08 | 4.607 | 129.99 | 61.10 | 0.68312 | 2650 | 688.90 | 556.3 | 507.25 | 1.765 | 1.01172 |
| 780 | 186.94 | 5.051 | 133.47 | 57.20 | 0.68942 | 2700 | 703.35 | 601.9 | 518.26 | 1.662 | 1.01712 |
| 800 | 191.81 | 5.526 | 136.97 | 53.63 | 0.69558 | 2750 | 717.83 | 650.4 | 529.31 | 1.566 | 1.02244 |
| 820 | 196.69 | 6.033 | 140.47 | 50.35 | 0.70160 | 2800 | 732.33 | 702.0 | 540.40 | 1.478 | 1.02767 |
| 840 | 201.56 | 6.573 | 143.98 | 47.34 | 0.70747 | 2850 | 746.88 | 756.7 | 551.52 | 1.395 | 1.03282 |
| 860 | 206.46 | 7.149 | 147.50 | 44.57 | 0.71323 | 2900 | 761.45 | 814.8 | 562.66 | 1.318 | 1.03788 |
| 880 | 211.35 | 7.761 | 151.02 | 42.01 | 0.71886 | 2950 | 776.05 | 876.4 | 573.84 | 1.247 | 1.04288 |
| 900 | 216.26 | 8.411 | 154.57 | 39.64 | 0.72438 | 3000 | 790.68 | 941.4 | 585.04 | 1.180 | 1.04779 |
| 920 | 221.18 | 9.102 | 158.12 | 37.44 | 0.72979 | 3050 | 805.34 | 1011 | 596.28 | 1.118 | 1.05264 |
| 940 | 226.11 | 9.834 | 161.68 | 35.41 | 0.73509 | 3100 | 820.03 | 1083 | 607.53 | 1.060 | 1.05741 |
| 960 | 231.06 | 10.61 | 165.26 | 33.52 | 0.74030 | 3150 | 834.75 | 1161 | 618.82 | 1.006 | 1.06212 |
| 980 | 236.02 | 11.43 | 168.83 | 31.76 | 0.74540 | 3200 | 849.48 | 1242 | 630.12 | 0.955 | 1.06676 |
| 1000 | 240.98 | 12.30 | 172.43 | 30.12 | 0.75042 | 3250 | 864.24 | 1328 | 641.46 | 0.907 | 1.07134 |
| 1040 | 250.95 | 14.18 | 179.66 | 27.17 | 0.76019 | 3300 | 879.02 | 1418 | 652.81 | 0.8621 | 1.07585 |
| 1080 | 260.97 | 16.28 | 186.93 | 24.58 | 0.76964 | 3350 | 893.83 | 1513 | 664.20 | 0.8202 | 1.08031 |
| 1120 | 271.03 | 18.60 | 194.25 | 22.30 | 0.77880 | 3400 | 908.66 | 1613 | 675.60 | 0.7807 | 1.08470 |
| 1160 | 281.14 | 21.18 | 201.63 | 20.29 | 0.78767 | 3450 | 923.52 | 1719 | 687.04 | 0.7436 | 1.08904 |
| 1200 | 291.30 | 24.01 | 209.05 | 18.51 | 0.79628 | 3500 | 938.40 | 1829 | 698.48 | 0.7087 | 1.09332 |
| 1240 | 301.52 | 27.13 | 216.53 | 16.93 | 0.80466 | 3550 | 953.30 | 1946 | 709.95 | 0.6759 | 1.09755 |
| 1280 | 311.79 | 30.55 | 224.05 | 15.52 | 0.81280 | 3600 | 968.21 | 2068 | 721.44 | 0.6449 | 1.10172 |
| 1320 | 322.11 | 34.31 | 231.63 | 14.25 | 0.82075 | 3650 | 983.15 | 2196 | 732.95 | 0.6157 | 1.10584 |
| 1360 | 332.48 | 38.41 | 239.25 | 13.12 | 0.82848 | 3700 | 998.11 | 2330 | 744.48 | 0.5882 | 1.10991 |
| 1400 | 342.90 | 42.88 | 246.93 | 12.10 | 0.83604 | 3750 | 1013.1 | 2471 | 756.04 | 0.5621 | 1.11393 |
| 1440 | 353.37 | 47.75 | 254.66 | 11.17 | 0.84341 | 3800 | 1028.1 | 2618 | 767.60 | 0.5376 | 1.11791 |
| 1480 | 363.89 | 53.04 | 262.44 | 10.34 | 0.85062 | 3850 | 1043.1 | 2773 | 779.19 | 0.5143 | 1.12183 |
| 1520 | 374.47 | 58.78 | 270.26 | 9.578 | 0.85767 | 3900 | 1058.1 | 2934 | 790.80 | 0.4923 | 1.12571 |
| 1560 | 385.08 | 65.00 | 278.13 | 8.890 | 0.86456 | 3950 | 1073.2 | 3103 | 802.43 | 0.4715 | 1.12955 |

## TABLE A-17E

Ideal-gas properties of air *(Concluded)*

| $T$ R | $h$ Btu/lbm | $P_r$ | $u$ Btu/lbm | $\upsilon_r$ | $s°$ Btu/lbm·R | $T$ R | $h$ Btu/lbm | $P_r$ | $u$ Btu/lbm | $\upsilon_r$ | $s°$ Btu/lbm·R |
|---|---|---|---|---|---|---|---|---|---|---|---|
| 4000 | 1088.3 | 3280 | 814.06 | 0.4518 | 1.13334 | 4600 | 1270.4 | 6089 | 955.04 | 0.2799 | 1.17575 |
| 4050 | 1103.4 | 3464 | 825.72 | 0.4331 | 1.13709 | 4700 | 1300.9 | 6701 | 978.73 | 0.2598 | 1.18232 |
| 4100 | 1118.5 | 3656 | 837.40 | 0.4154 | 1.14079 | 4800 | 1331.5 | 7362 | 1002.5 | 0.2415 | 1.18876 |
| 4150 | 1133.6 | 3858 | 849.09 | 0.3985 | 1.14446 | 4900 | 1362.2 | 8073 | 1026.3 | 0.2248 | 1.19508 |
| 4200 | 1148.7 | 4067 | 860.81 | 0.3826 | 1.14809 | 5000 | 1392.9 | 8837 | 1050.1 | 0.2096 | 1.20129 |
| 4300 | 1179.0 | 4513 | 884.28 | 0.3529 | 1.15522 | 5100 | 1423.6 | 9658 | 1074.0 | 0.1956 | 1.20738 |
| 4400 | 1209.4 | 4997 | 907.81 | 0.3262 | 1.16221 | 5200 | 1454.4 | 10,539 | 1098.0 | 0.1828 | 1.21336 |
| 4500 | 1239.9 | 5521 | 931.39 | 0.3019 | 1.16905 | 5300 | 1485.3 | 11,481 | 1122.0 | 0.1710 | 1.2192 |

*Note:* The properties $P_r$ (relative pressure) and $\upsilon_r$ (relative specific volume) are dimensionless quantities used in the analysis of isentropic processes, and should not be confused with the properties pressure and specific volume.

*Source of Data:* Kenneth Wark, *Thermodynamics,* 4th ed. (New York: McGraw-Hill, 1983), pp. 832–33, Table A–5. Originally published in J. H. Keenan and J. Kaye, *Gas Tables* (New York: John Wiley & Sons, 1948).

## TABLE A–18E

Ideal-gas properties of nitrogen, $N_2$

| $T$ R | $\bar{h}$ Btu/lbmol | $\bar{u}$ Btu/lbmol | $\bar{s}°$ Btu/lbmol·R | $T$ R | $\bar{h}$ Btu/lbmol | $\bar{u}$ Btu/lbmol | $\bar{s}°$ Btu/lbmol·R |
|---|---|---|---|---|---|---|---|
| 300 | 2,082.0 | 1,486.2 | 41.695 | 1080 | 7,551.0 | 5,406.2 | 50.651 |
| 320 | 2,221.0 | 1,585.5 | 42.143 | 1100 | 7,695.0 | 5,510.5 | 50.783 |
| 340 | 2,360.0 | 1,684.4 | 42.564 | 1120 | 7,839.3 | 5,615.2 | 50.912 |
| 360 | 2,498.9 | 1,784.0 | 42.962 | 1140 | 7,984.0 | 5,720.1 | 51.040 |
| 380 | 2,638.0 | 1,883.4 | 43.337 | 1160 | 8,129.0 | 5,825.4 | 51.167 |
| 400 | 2,777.0 | 1,982.6 | 43.694 | 1180 | 8,274.4 | 5,931.0 | 51.291 |
| 420 | 2,916.1 | 2,082.0 | 44.034 | 1200 | 8,420.0 | 6,037.0 | 51.143 |
| 440 | 3,055.1 | 2,181.3 | 44.357 | 1220 | 8,566.1 | 6,143.4 | 51.534 |
| 460 | 3,194.1 | 2,280.6 | 44.665 | 1240 | 8,712.6 | 6,250.1 | 51.653 |
| 480 | 3,333.1 | 2,379.9 | 44.962 | 1260 | 8,859.3 | 6,357.2 | 51.771 |
| 500 | 3,472.2 | 2,479.3 | 45.246 | 1280 | 9,006.4 | 6,464.5 | 51.887 |
| 520 | 3,611.3 | 2,578.6 | 45.519 | 1300 | 9,153.9 | 6,572.3 | 51.001 |
| 537 | 3,729.5 | 2,663.1 | 45.743 | 1320 | 9,301.8 | 6,680.4 | 52.114 |
| 540 | 3,750.3 | 2,678.0 | 45.781 | 1340 | 9,450.0 | 6,788.9 | 52.225 |
| 560 | 3,889.5 | 2,777.4 | 46.034 | 1360 | 9,598.6 | 6,897.8 | 52.335 |
| 580 | 4,028.7 | 2,876.9 | 46.278 | 1380 | 9,747.5 | 7,007.0 | 52.444 |
| 600 | 4,167.9 | 2,976.4 | 46.514 | 1400 | 9,896.9 | 7,116.7 | 52.551 |
| 620 | 4,307.1 | 3,075.9 | 46.742 | 1420 | 10,046.6 | 7,226.7 | 52.658 |
| 640 | 4,446.4 | 3,175.5 | 46.964 | 1440 | 10,196.6 | 7,337.0 | 52.763 |
| 660 | 4,585.8 | 3,275.2 | 47.178 | 1460 | 10,347.0 | 7,447.6 | 52.867 |
| 680 | 4,725.3 | 3,374.9 | 47.386 | 1480 | 10,497.8 | 7,558.7 | 52.969 |
| 700 | 4,864.9 | 3,474.8 | 47.588 | 1500 | 10,648.0 | 7,670.1 | 53.071 |
| 720 | 5,004.5 | 3,574.7 | 47.785 | 1520 | 10,800.4 | 7,781.9 | 53.171 |
| 740 | 5,144.3 | 3,674.7 | 47.977 | 1540 | 10,952.2 | 7,893.9 | 53.271 |
| 760 | 5,284.1 | 3,774.9 | 48.164 | 1560 | 11,104.3 | 8,006.4 | 53.369 |
| 780 | 5,424.2 | 3,875.2 | 48.345 | 1580 | 11,256.9 | 8,119.2 | 53.465 |
| 800 | 5,564.4 | 3,975.7 | 48.522 | 1600 | 11,409.7 | 8,232.3 | 53.561 |
| 820 | 5,704.7 | 4,076.3 | 48.696 | 1620 | 11,562.8 | 8,345.7 | 53.656 |
| 840 | 5,845.3 | 4,177.1 | 48.865 | 1640 | 11,716.4 | 8,459.6 | 53.751 |
| 860 | 5,985.9 | 4,278.1 | 49.031 | 1660 | 11,870.2 | 8,573.6 | 53.844 |
| 880 | 6,126.9 | 4,379.4 | 49.193 | 1680 | 12,024.3 | 8,688.1 | 53.936 |
| 900 | 6,268.1 | 4,480.8 | 49.352 | 1700 | 12,178.9 | 8,802.9 | 54.028 |
| 920 | 6,409.6 | 4,582.6 | 49.507 | 1720 | 12,333.7 | 8,918.0 | 54.118 |
| 940 | 6,551.2 | 4,684.5 | 49.659 | 1740 | 12,488.8 | 9,033.4 | 54.208 |
| 960 | 6,693.1 | 4,786.7 | 49.808 | 1760 | 12,644.3 | 9,149.2 | 54.297 |
| 980 | 6,835.4 | 4,889.3 | 49.955 | 1780 | 12,800.2 | 9,265.3 | 54.385 |
| 1000 | 6,977.9 | 4,992.0 | 50.099 | 1800 | 12,956.3 | 9,381.7 | 54.472 |
| 1020 | 7,120.7 | 5,095.1 | 50.241 | 1820 | 13,112.7 | 9,498.4 | 54.559 |
| 1040 | 7,263.8 | 5,198.5 | 50.380 | 1840 | 13,269.5 | 9,615.5 | 54.645 |
| 1060 | 7,407.2 | 5,302.2 | 50.516 | 1860 | 13,426.5 | 9,732.8 | 54.729 |

Ideal-gas properties of nitrogen, $N_2$ *(Concluded)*

| $T$ R | $\bar{h}$ Btu/lbmol | $\bar{u}$ Btu/lbmol | $\bar{s}°$ Btu/lbmol·R | $T$ R | $\bar{h}$ Btu/lbmol | $\bar{u}$ Btu/lbmol | $\bar{s}°$ Btu/lbmol·R |
|---|---|---|---|---|---|---|---|
| 1900 | 13,742 | 9,968 | 54.896 | 3500 | 27,016 | 20,065 | 59.944 |
| 1940 | 14,058 | 10,205 | 55.061 | 3540 | 27,359 | 20,329 | 60.041 |
| 1980 | 14,375 | 10,443 | 55.223 | 3580 | 27,703 | 20,593 | 60.138 |
| 2020 | 14,694 | 10,682 | 55.383 | 3620 | 28,046 | 20,858 | 60.234 |
| 2060 | 15,013 | 10,923 | 55.540 | 3660 | 28,391 | 21,122 | 60.328 |
| 2100 | 15,334 | 11,164 | 55.694 | 3700 | 28,735 | 21,387 | 60.422 |
| 2140 | 15,656 | 11,406 | 55.846 | 3740 | 29,080 | 21,653 | 60.515 |
| 2180 | 15,978 | 11,649 | 55.995 | 3780 | 29,425 | 21,919 | 60.607 |
| 2220 | 16,302 | 11,893 | 56.141 | 3820 | 29,771 | 22,185 | 60.698 |
| 2260 | 16,626 | 12,138 | 56.286 | 3860 | 30,117 | 22,451 | 60.788 |
| 2300 | 16,951 | 12,384 | 56.429 | 3900 | 30,463 | 22,718 | 60.877 |
| 2340 | 17,277 | 12,630 | 56.570 | 3940 | 30,809 | 22,985 | 60.966 |
| 2380 | 17,604 | 12,878 | 56.708 | 3980 | 31,156 | 23,252 | 61.053 |
| 2420 | 17,392 | 13,126 | 56.845 | 4020 | 31,503 | 23,520 | 61.139 |
| 2460 | 18,260 | 13,375 | 56.980 | 4060 | 31,850 | 23,788 | 61.225 |
| 2500 | 18,590 | 13,625 | 57.112 | 4100 | 32,198 | 24,056 | 61.310 |
| 2540 | 18,919 | 13,875 | 57.243 | 4140 | 32,546 | 24,324 | 61.395 |
| 2580 | 19,250 | 14,127 | 57.372 | 4180 | 32,894 | 24,593 | 61.479 |
| 2620 | 19,582 | 14,379 | 57.499 | 4220 | 33,242 | 24,862 | 61.562 |
| 2660 | 19,914 | 14,631 | 57.625 | 4260 | 33,591 | 25,131 | 61.644 |
| 2700 | 20,246 | 14,885 | 57.750 | 4300 | 33,940 | 25,401 | 61.726 |
| 2740 | 20,580 | 15,139 | 57.872 | 4340 | 34,289 | 25,670 | 61.806 |
| 2780 | 20,914 | 15,393 | 57.993 | 4380 | 34,638 | 25,940 | 61.887 |
| 2820 | 21,248 | 15,648 | 58.113 | 4420 | 34,988 | 26,210 | 61.966 |
| 2860 | 21,584 | 15,905 | 58.231 | 4460 | 35,338 | 26,481 | 62.045 |
| 2900 | 21,920 | 16,161 | 58.348 | 4500 | 35,688 | 26,751 | 62.123 |
| 2940 | 22,256 | 16,417 | 58.463 | 4540 | 36,038 | 27,022 | 62.201 |
| 2980 | 22,593 | 16,675 | 58.576 | 4580 | 36,389 | 27,293 | 62.278 |
| 3020 | 22,930 | 16,933 | 58.688 | 4620 | 36,739 | 27,565 | 62.354 |
| 3060 | 23,268 | 17,192 | 58.800 | 4660 | 37,090 | 27,836 | 62.429 |
| 3100 | 23,607 | 17,451 | 58.910 | 4700 | 37,441 | 28,108 | 62.504 |
| 3140 | 23,946 | 17,710 | 59.019 | 4740 | 37,792 | 28,379 | 62.578 |
| 3180 | 24,285 | 17,970 | 59.126 | 4780 | 38,144 | 28,651 | 62.652 |
| 3220 | 24,625 | 18,231 | 59.232 | 4820 | 38,495 | 28,924 | 62.725 |
| 3260 | 24,965 | 18,491 | 59.338 | 4860 | 38,847 | 29,196 | 62.798 |
| 3300 | 25,306 | 18,753 | 59.442 | 4900 | 39,199 | 29,468 | 62.870 |
| 3340 | 25,647 | 19,014 | 59.544 | 5000 | 40,080 | 30,151 | 63.049 |
| 3380 | 25,989 | 19,277 | 59.646 | 5100 | 40,962 | 30,834 | 63.223 |
| 3420 | 26,331 | 19,539 | 59.747 | 5200 | 41,844 | 31,518 | 63.395 |
| 3460 | 26,673 | 19,802 | 59.846 | 5300 | 42,728 | 32,203 | 63.563 |

*Source of Data:* Tables A–18E through A–23E are adapted from Kenneth Wark, *Thermodynamics,* 4th ed. (New York: McGraw-Hill, 1983), pp. 834–44.
Originally published in J. H. Keenan and J. Kaye, *Gas Tables* (New York: John Wiley & Sons, 1945).

## TABLE A–19E

Ideal-gas properties of oxygen, $O_2$

| $T$ R | $\bar{h}$ Btu/lbmol | $\bar{u}$ Btu/lbmol | $\bar{s}°$ Btu/lbmol·R | $T$ R | $\bar{h}$ Btu/lbmol | $\bar{u}$ Btu/lbmol | $\bar{s}°$ Btu/lbmol·R |
|---|---|---|---|---|---|---|---|
| 300 | 2,073.5 | 1,477.8 | 44.927 | 1080 | 7,696.8 | 5,552.1 | 54.064 |
| 320 | 2,212.6 | 1,577.1 | 45.375 | 1100 | 7,850.4 | 5,665.9 | 54.204 |
| 340 | 2,351.7 | 1,676.5 | 45.797 | 1120 | 8,004.5 | 5,780.3 | 54.343 |
| 360 | 2,490.8 | 1,775.9 | 46.195 | 1140 | 8,159.1 | 5,895.2 | 54.480 |
| 380 | 2,630.0 | 1,875.3 | 46.571 | 1160 | 8,314.2 | 6,010.6 | 54.614 |
| 400 | 2,769.1 | 1,974.8 | 46.927 | 1180 | 8,469.8 | 6,126.5 | 54.748 |
| 420 | 2,908.3 | 2,074.3 | 47.267 | 1200 | 8,625.8 | 6,242.8 | 54.879 |
| 440 | 3,047.5 | 2,173.8 | 47.591 | 1220 | 8,782.4 | 6,359.6 | 55.008 |
| 460 | 3,186.9 | 2,273.4 | 47.900 | 1240 | 8,939.4 | 6,476.9 | 55.136 |
| 480 | 3,326.5 | 2,373.3 | 48.198 | 1260 | 9,096.7 | 6,594.5 | 55.262 |
| 500 | 3,466.2 | 2,473.2 | 48.483 | 1280 | 9,254.6 | 6,712.7 | 55.386 |
| 520 | 3,606.1 | 2,573.4 | 48.757 | 1300 | 9,412.9 | 6,831.3 | 55.508 |
| 537 | 3,725.1 | 2,658.7 | 48.982 | 1320 | 9,571.9 | 6,950.2 | 55.630 |
| 540 | 3,746.2 | 2,673.8 | 49.021 | 1340 | 9,730.7 | 7,069.6 | 55.750 |
| 560 | 3,886.6 | 2,774.5 | 49.276 | 1360 | 9,890.2 | 7,189.4 | 55.867 |
| 580 | 4,027.3 | 2,875.5 | 49.522 | 1380 | 10,050.1 | 7,309.6 | 55.984 |
| 600 | 4,168.3 | 2,976.8 | 49.762 | 1400 | 10,210.4 | 7,430.1 | 56.099 |
| 620 | 4,309.7 | 3,078.4 | 49.993 | 1420 | 10,371.0 | 7,551.1 | 56.213 |
| 640 | 4,451.4 | 3,180.4 | 50.218 | 1440 | 10,532.0 | 7,672.4 | 56.326 |
| 660 | 4,593.5 | 3,282.9 | 50.437 | 1460 | 10,693.3 | 7,793.9 | 56.437 |
| 680 | 4,736.2 | 3,385.8 | 50.650 | 1480 | 10,855.1 | 7,916.0 | 56.547 |
| 700 | 4,879.3 | 3,489.2 | 50.858 | 1500 | 11,017.1 | 8,038.3 | 56.656 |
| 720 | 5,022.9 | 3,593.1 | 51.059 | 1520 | 11,179.6 | 8,161.1 | 56.763 |
| 740 | 5,167.0 | 3,697.4 | 51.257 | 1540 | 11,342.4 | 8,284.2 | 56.869 |
| 760 | 5,311.4 | 3,802.4 | 51.450 | 1560 | 11,505.4 | 8,407.4 | 56.975 |
| 780 | 5,456.4 | 3,907.5 | 51.638 | 1580 | 11,668.8 | 8,531.1 | 57.079 |
| 800 | 5,602.0 | 4,013.3 | 51.821 | 1600 | 11,832.5 | 8,655.1 | 57.182 |
| 820 | 5,748.1 | 4,119.7 | 52.002 | 1620 | 11,996.6 | 8,779.5 | 57.284 |
| 840 | 5,894.8 | 4,226.6 | 52.179 | 1640 | 12,160.9 | 8,904.1 | 57.385 |
| 860 | 6,041.9 | 4,334.1 | 52.352 | 1660 | 12,325.5 | 9,029.0 | 57.484 |
| 880 | 6,189.6 | 4,442.0 | 52.522 | 1680 | 12,490.4 | 9,154.1 | 57.582 |
| 900 | 6,337.9 | 4,550.6 | 52.688 | 1700 | 12,655.6 | 9,279.6 | 57.680 |
| 920 | 6,486.7 | 4,659.7 | 52.852 | 1720 | 12,821.1 | 9,405.4 | 57.777 |
| 940 | 6,636.1 | 4,769.4 | 53.012 | 1740 | 12,986.9 | 9,531.5 | 57.873 |
| 960 | 6,786.0 | 4,879.5 | 53.170 | 1760 | 13,153.0 | 9,657.9 | 57.968 |
| 980 | 6,936.4 | 4,990.3 | 53.326 | 1780 | 13,319.2 | 9,784.4 | 58.062 |
| 1000 | 7,087.5 | 5,101.6 | 53.477 | 1800 | 13,485.8 | 9,911.2 | 58.155 |
| 1020 | 7,238.9 | 5,213.3 | 53.628 | 1820 | 13,652.5 | 10,038.2 | 58.247 |
| 1040 | 7,391.0 | 5,325.7 | 53.775 | 1840 | 13,819.6 | 10,165.6 | 58.339 |
| 1060 | 7,543.6 | 5,438.6 | 53.921 | 1860 | 13,986.8 | 10,293.1 | 58.428 |

## TABLE A–19E

Ideal-gas properties of oxygen, $O_2$ *(Concluded)*

| T R | $\bar{h}$ Btu/lbmol | $\bar{u}$ Btu/lbmol | $\bar{s}°$ Btu/lbmol·R | T R | $\bar{h}$ Btu/lbmol | $\bar{u}$ Btu/lbmol | $\bar{s}°$ Btu/lbmol·R |
|---|---|---|---|---|---|---|---|
| 1900 | 14,322 | 10,549 | 58.607 | 3500 | 28,273 | 21,323 | 63.914 |
| 1940 | 14,658 | 10,806 | 58.782 | 3540 | 28,633 | 21,603 | 64.016 |
| 1980 | 14,995 | 11,063 | 58.954 | 3580 | 28,994 | 21,884 | 64.114 |
| 2020 | 15,333 | 11,321 | 59.123 | 3620 | 29,354 | 22,165 | 64.217 |
| 2060 | 15,672 | 11,581 | 59.289 | 3660 | 29,716 | 22,447 | 64.316 |
| 2100 | 16,011 | 11,841 | 59.451 | 3700 | 30,078 | 22,730 | 64.415 |
| 2140 | 16,351 | 12,101 | 59.612 | 3740 | 30,440 | 23,013 | 64.512 |
| 2180 | 16,692 | 12,363 | 59.770 | 3780 | 30,803 | 23,296 | 64.609 |
| 2220 | 17,036 | 12,625 | 59.926 | 3820 | 31,166 | 23,580 | 64.704 |
| 2260 | 17,376 | 12,888 | 60.077 | 3860 | 31,529 | 23,864 | 64.800 |
| 2300 | 17,719 | 13,151 | 60.228 | 3900 | 31,894 | 24,149 | 64.893 |
| 2340 | 18,062 | 13,416 | 60.376 | 3940 | 32,258 | 24,434 | 64.986 |
| 2380 | 18,407 | 13,680 | 60.522 | 3980 | 32,623 | 24,720 | 65.078 |
| 2420 | 18,572 | 13,946 | 60.666 | 4020 | 32,989 | 25,006 | 65.169 |
| 2460 | 19,097 | 14,212 | 60.808 | 4060 | 33,355 | 25,292 | 65.260 |
| 2500 | 19,443 | 14,479 | 60.946 | 4100 | 33,722 | 25,580 | 65.350 |
| 2540 | 19,790 | 14,746 | 61.084 | 4140 | 34,089 | 25,867 | 64.439 |
| 2580 | 20,138 | 15,014 | 61.220 | 4180 | 34,456 | 26,155 | 65.527 |
| 2620 | 20,485 | 15,282 | 61.354 | 4220 | 34,824 | 26,144 | 65.615 |
| 2660 | 20,834 | 15,551 | 61.486 | 4260 | 35,192 | 26,733 | 65.702 |
| 2700 | 21,183 | 15,821 | 61.616 | 4300 | 35,561 | 27,022 | 65.788 |
| 2740 | 21,533 | 16,091 | 61.744 | 4340 | 35,930 | 27,312 | 65.873 |
| 2780 | 21,883 | 16,362 | 61.871 | 4380 | 36,300 | 27,602 | 65.958 |
| 2820 | 22,232 | 16,633 | 61.996 | 4420 | 36,670 | 27,823 | 66.042 |
| 2860 | 22,584 | 16,905 | 62.120 | 4460 | 37,041 | 28,184 | 66.125 |
| 2900 | 22,936 | 17,177 | 62.242 | 4500 | 37,412 | 28,475 | 66.208 |
| 2940 | 23,288 | 17,450 | 62.363 | 4540 | 37,783 | 28,768 | 66.290 |
| 2980 | 23,641 | 17,723 | 62.483 | 4580 | 38,155 | 29,060 | 66.372 |
| 3020 | 23,994 | 17,997 | 62.599 | 4620 | 38,528 | 29,353 | 66.453 |
| 3060 | 24,348 | 18,271 | 62.716 | 4660 | 38,900 | 29,646 | 66.533 |
| 3100 | 24,703 | 18,546 | 62.831 | 4700 | 39,274 | 29,940 | 66.613 |
| 3140 | 25,057 | 18,822 | 62.945 | 4740 | 39,647 | 30,234 | 66.691 |
| 3180 | 25,413 | 19,098 | 63.057 | 4780 | 40,021 | 30,529 | 66.770 |
| 3220 | 25,769 | 19,374 | 63.169 | 4820 | 40,396 | 30,824 | 66.848 |
| 3260 | 26,175 | 19,651 | 63.279 | 4860 | 40,771 | 31,120 | 66.925 |
| 3300 | 26,412 | 19,928 | 63.386 | 4900 | 41,146 | 31,415 | 67.003 |
| 3340 | 26,839 | 20,206 | 63.494 | 5000 | 42,086 | 32,157 | 67.193 |
| 3380 | 27,197 | 20,485 | 63.601 | 5100 | 43,021 | 32,901 | 67.380 |
| 3420 | 27,555 | 20,763 | 63.706 | 5200 | 43,974 | 33,648 | 67.562 |
| 3460 | 27,914 | 21,043 | 63.811 | 5300 | 44,922 | 34,397 | 67.743 |

Ideal-gas properties of carbon dioxide, $CO_2$

| $T$ R | $\bar{h}$ Btu/lbmol | $\bar{u}$ Btu/lbmol | $\bar{s}°$ Btu/lbmol·R | $T$ R | $\bar{h}$ Btu/lbmol | $\bar{u}$ Btu/lbmol | $\bar{s}°$ Btu/lbmol·R |
|---|---|---|---|---|---|---|---|
| 300 | 2,108.2 | 1,512.4 | 46.353 | 1080 | 9,575.8 | 7,431.1 | 58.072 |
| 320 | 2,256.6 | 1,621.1 | 46.832 | 1100 | 9,802.6 | 7,618.1 | 58.281 |
| 340 | 2,407.3 | 1,732.1 | 47.289 | 1120 | 10,030.6 | 7,806.4 | 58.485 |
| 360 | 2,560.5 | 1,845.6 | 47.728 | 1140 | 10,260.1 | 7,996.2 | 58.689 |
| 380 | 2,716.4 | 1,961.8 | 48.148 | 1160 | 10,490.6 | 8,187.0 | 58.889 |
| 400 | 2,874.7 | 2,080.4 | 48.555 | 1180 | 10,722.3 | 8,379.0 | 59.088 |
| 420 | 3,035.7 | 2,201.7 | 48.947 | 1200 | 10,955.3 | 8,572.3 | 59.283 |
| 440 | 3,199.4 | 2,325.6 | 49.329 | 1220 | 11,189.4 | 8,766.6 | 59.477 |
| 460 | 3,365.7 | 2,452.2 | 49.698 | 1240 | 11,424.6 | 8,962.1 | 59.668 |
| 480 | 3,534.7 | 2,581.5 | 50.058 | 1260 | 11,661.0 | 9,158.8 | 59.858 |
| 500 | 3,706.2 | 2,713.3 | 50.408 | 1280 | 11,898.4 | 9,356.5 | 60.044 |
| 520 | 3,880.3 | 2,847.7 | 50.750 | 1300 | 12,136.9 | 9,555.3 | 60.229 |
| 537 | 4,027.5 | 2,963.8 | 51.032 | 1320 | 12,376.4 | 9,755.0 | 60.412 |
| 540 | 4,056.8 | 2,984.4 | 51.082 | 1340 | 12,617.0 | 9,955.9 | 60.593 |
| 560 | 4,235.8 | 3,123.7 | 51.408 | 1360 | 12,858.5 | 10,157.7 | 60.772 |
| 580 | 4,417.2 | 3,265.4 | 51.726 | 1380 | 13,101.0 | 10,360.5 | 60.949 |
| 600 | 4,600.9 | 3,409.4 | 52.038 | 1400 | 13,344.7 | 10,564.5 | 61.124 |
| 620 | 4,786.6 | 3,555.6 | 52.343 | 1420 | 13,589.1 | 10,769.2 | 61.298 |
| 640 | 4,974.9 | 3,704.0 | 52.641 | 1440 | 13,834.5 | 10,974.8 | 61.469 |
| 660 | 5,165.2 | 3,854.6 | 52.934 | 1460 | 14,080.8 | 11,181.4 | 61.639 |
| 680 | 5,357.6 | 4,007.2 | 53.225 | 1480 | 14,328.0 | 11,388.9 | 61.800 |
| 700 | 5,552.0 | 4,161.9 | 53.503 | 1500 | 14,576.0 | 11,597.2 | 61.974 |
| 720 | 5,748.4 | 4,318.6 | 53.780 | 1520 | 14,824.9 | 11,806.4 | 62.138 |
| 740 | 5,946.8 | 4,477.3 | 54.051 | 1540 | 15,074.7 | 12,016.5 | 62.302 |
| 760 | 6,147.0 | 4,637.9 | 54.319 | 1560 | 15,325.3 | 12,227.3 | 62.464 |
| 780 | 6,349.1 | 4,800.1 | 54.582 | 1580 | 15,576.7 | 12,439.0 | 62.624 |
| 800 | 6,552.9 | 4,964.2 | 54.839 | 1600 | 15,829.0 | 12,651.6 | 62.783 |
| 820 | 6,758.3 | 5,129.9 | 55.093 | 1620 | 16,081.9 | 12,864.8 | 62.939 |
| 840 | 6,965.7 | 5,297.6 | 55.343 | 1640 | 16,335.7 | 13,078.9 | 63.095 |
| 860 | 7,174.7 | 5,466.9 | 55.589 | 1660 | 16,590.2 | 13,293.7 | 63.250 |
| 880 | 7,385.3 | 5,637.7 | 55.831 | 1680 | 16,845.5 | 13,509.2 | 63.403 |
| 900 | 7,597.6 | 5,810.3 | 56.070 | 1700 | 17,101.4 | 13,725.4 | 63.555 |
| 920 | 7,811.4 | 5,984.4 | 56.305 | 1720 | 17,358.1 | 13,942.4 | 63.704 |
| 940 | 8,026.8 | 6,160.1 | 56.536 | 1740 | 17,615.5 | 14,160.1 | 63.853 |
| 960 | 8,243.8 | 6,337.4 | 56.765 | 1760 | 17,873.5 | 14,378.4 | 64.001 |
| 980 | 8,462.2 | 6,516.1 | 56.990 | 1780 | 18,132.2 | 14,597.4 | 64.147 |
| 1000 | 8,682.1 | 6,696.2 | 57.212 | 1800 | 18,391.5 | 14,816.9 | 64.292 |
| 1020 | 8,903.4 | 6,877.8 | 57.432 | 1820 | 18,651.5 | 15,037.2 | 64.435 |
| 1040 | 9,126.2 | 7,060.9 | 57.647 | 1840 | 18,912.2 | 15,258.2 | 64.578 |
| 1060 | 9,350.3 | 7,245.3 | 57.861 | 1860 | 19,173.4 | 15,479.7 | 64.719 |

Ideal-gas properties of carbon dioxide, $CO_2$ *(Concluded)*

| $T$ R | $\bar{h}$ Btu/lbmol | $\bar{u}$ Btu/lbmol | $\bar{s}°$ Btu/lbmol·R | $T$ R | $\bar{h}$ Btu/lbmol | $\bar{u}$ Btu/lbmol | $\bar{s}°$ Btu/lbmol·R |
|---|---|---|---|---|---|---|---|
| 1900 | 19,698 | 15,925 | 64.999 | 3500 | 41,965 | 35,015 | 73.462 |
| 1940 | 20,224 | 16,372 | 65.272 | 3540 | 42,543 | 35,513 | 73.627 |
| 1980 | 20,753 | 16,821 | 65.543 | 3580 | 43,121 | 36,012 | 73.789 |
| 2020 | 21,284 | 17,273 | 65.809 | 3620 | 43,701 | 36,512 | 73.951 |
| 2060 | 21,818 | 17,727 | 66.069 | 3660 | 44,280 | 37,012 | 74.110 |
| 2100 | 22,353 | 18,182 | 66.327 | 3700 | 44,861 | 37,513 | 74.267 |
| 2140 | 22,890 | 18,640 | 66.581 | 3740 | 45,442 | 38,014 | 74.423 |
| 2180 | 23,429 | 19,101 | 66.830 | 3780 | 46,023 | 38,517 | 74.578 |
| 2220 | 23,970 | 19,561 | 67.076 | 3820 | 46,605 | 39,019 | 74.732 |
| 2260 | 24,512 | 20,024 | 67.319 | 3860 | 47,188 | 39,522 | 74.884 |
| 2300 | 25,056 | 20,489 | 67.557 | 3900 | 47,771 | 40,026 | 75.033 |
| 2340 | 25,602 | 20,955 | 67.792 | 3940 | 48,355 | 40,531 | 75.182 |
| 2380 | 26,150 | 21,423 | 68.025 | 3980 | 48,939 | 41,035 | 75.330 |
| 2420 | 26,699 | 21,893 | 68.253 | 4020 | 49,524 | 41,541 | 75.477 |
| 2460 | 27,249 | 22,364 | 68.479 | 4060 | 50,109 | 42,047 | 75.622 |
| 2500 | 27,801 | 22,837 | 68.702 | 4100 | 50,695 | 42,553 | 75.765 |
| 2540 | 28,355 | 23,310 | 68.921 | 4140 | 51,282 | 43,060 | 75.907 |
| 2580 | 28,910 | 23,786 | 69.138 | 4180 | 51,868 | 43,568 | 76.048 |
| 2620 | 29,465 | 24,262 | 69.352 | 4220 | 52,456 | 44,075 | 76.188 |
| 2660 | 30,023 | 24,740 | 69.563 | 4260 | 53,044 | 44,584 | 76.327 |
| 2700 | 30,581 | 25,220 | 69.771 | 4300 | 53,632 | 45,093 | 76.464 |
| 2740 | 31,141 | 25,701 | 69.977 | 4340 | 54,221 | 45,602 | 76.601 |
| 2780 | 31,702 | 26,181 | 70.181 | 4380 | 54,810 | 46,112 | 76.736 |
| 2820 | 32,264 | 26,664 | 70.382 | 4420 | 55,400 | 46,622 | 76.870 |
| 2860 | 32,827 | 27,148 | 70.580 | 4460 | 55,990 | 47,133 | 77.003 |
| 2900 | 33,392 | 27,633 | 70.776 | 4500 | 56,581 | 47,645 | 77.135 |
| 2940 | 33,957 | 28,118 | 70.970 | 4540 | 57,172 | 48,156 | 77.266 |
| 2980 | 34,523 | 28,605 | 71.160 | 4580 | 57,764 | 48,668 | 77.395 |
| 3020 | 35,090 | 29,093 | 71.350 | 4620 | 58,356 | 49,181 | 77.581 |
| 3060 | 35,659 | 29,582 | 71.537 | 4660 | 58,948 | 49,694 | 77.652 |
| 3100 | 36,228 | 30,072 | 71.722 | 4700 | 59,541 | 50,208 | 77.779 |
| 3140 | 36,798 | 30,562 | 71.904 | 4740 | 60,134 | 50,721 | 77.905 |
| 3180 | 37,369 | 31,054 | 72.085 | 4780 | 60,728 | 51,236 | 78.029 |
| 3220 | 37,941 | 31,546 | 72.264 | 4820 | 61,322 | 51,750 | 78.153 |
| 3260 | 38,513 | 32,039 | 72.441 | 4860 | 61,916 | 52,265 | 78.276 |
| 3300 | 39,087 | 32,533 | 72.616 | 4900 | 62,511 | 52,781 | 78.398 |
| 3340 | 39,661 | 33,028 | 72.788 | 5000 | 64,000 | 54,071 | 78.698 |
| 3380 | 40,236 | 33,524 | 72.960 | 5100 | 65,491 | 55,363 | 78.994 |
| 3420 | 40,812 | 34,020 | 73.129 | 5200 | 66,984 | 56,658 | 79.284 |
| 3460 | 41,388 | 34,517 | 73.297 | 5300 | 68,471 | 57,954 | 79.569 |

Ideal-gas properties of carbon monoxide, CO

| T R | $\bar{h}$ Btu/lbmol | $\bar{u}$ Btu/lbmol | $\bar{s}°$ Btu/lbmol·R | T R | $\bar{h}$ Btu/lbmol | $\bar{u}$ Btu/lbmol | $\bar{s}°$ Btu/lbmol·R |
|---|---|---|---|---|---|---|---|
| 300 | 2,081.9 | 1,486.1 | 43.223 | 1080 | 7,571.1 | 5,426.4 | 52.203 |
| 320 | 2,220.9 | 1,585.4 | 43.672 | 1100 | 7,716.8 | 5,532.3 | 52.337 |
| 340 | 2,359.9 | 1,684.7 | 44.093 | 1120 | 7,862.9 | 5,638.7 | 52.468 |
| 360 | 2,498.8 | 1,783.9 | 44.490 | 1140 | 8,009.2 | 5,745.4 | 52.598 |
| 380 | 2,637.9 | 1,883.3 | 44.866 | 1160 | 8,156.1 | 5,851.5 | 52.726 |
| 400 | 2,776.9 | 1,982.6 | 45.223 | 1180 | 8,303.3 | 5,960.0 | 52.852 |
| 420 | 2,916.0 | 2,081.9 | 45.563 | 1200 | 8,450.8 | 6,067.8 | 52.976 |
| 440 | 3,055.0 | 2,181.2 | 45.886 | 1220 | 8,598.8 | 6,176.0 | 53.098 |
| 460 | 3,194.0 | 2,280.5 | 46.194 | 1240 | 8,747.2 | 6,284.7 | 53.218 |
| 480 | 3,333.0 | 2,379.8 | 46.491 | 1260 | 8,896.0 | 6,393.8 | 53.337 |
| 500 | 3,472.1 | 2,479.2 | 46.775 | 1280 | 9,045.0 | 6,503.1 | 53.455 |
| 520 | 3,611.2 | 2,578.6 | 47.048 | 1300 | 9,194.6 | 6,613.0 | 53.571 |
| 537 | 3,725.1 | 2,663.1 | 47.272 | 1320 | 9,344.6 | 6,723.2 | 53.685 |
| 540 | 3,750.3 | 2,677.9 | 47.310 | 1340 | 9,494.8 | 6,833.7 | 53.799 |
| 560 | 3,889.5 | 2,777.4 | 47.563 | 1360 | 9,645.5 | 6,944.7 | 53.910 |
| 580 | 4,028.7 | 2,876.9 | 47.807 | 1380 | 9,796.6 | 7,056.1 | 54.021 |
| 600 | 4,168.0 | 2,976.5 | 48.044 | 1400 | 9,948.1 | 7,167.9 | 54.129 |
| 620 | 4,307.4 | 3,076.2 | 48.272 | 1420 | 10,100.0 | 7,280.1 | 54.237 |
| 640 | 4,446.9 | 3,175.9 | 48.494 | 1440 | 10,252.2 | 7,392.6 | 54.344 |
| 660 | 4,586.6 | 3,275.8 | 48.709 | 1460 | 10,404.8 | 7,505.4 | 54.448 |
| 680 | 4,726.2 | 3,375.8 | 48.917 | 1480 | 10,557.8 | 7,618.7 | 54.522 |
| 700 | 4,886.0 | 3,475.9 | 49.120 | 1500 | 10,711.1 | 7,732.3 | 54.665 |
| 720 | 5,006.1 | 3,576.3 | 49.317 | 1520 | 10,864.9 | 7,846.4 | 54.757 |
| 740 | 5,146.4 | 3,676.9 | 49.509 | 1540 | 11,019.0 | 7,960.8 | 54.858 |
| 760 | 5,286.8 | 3,777.5 | 49.697 | 1560 | 11,173.4 | 8,075.4 | 54.958 |
| 780 | 5,427.4 | 3,878.4 | 49.880 | 1580 | 11,328.2 | 8,190.5 | 55.056 |
| 800 | 5,568.2 | 3,979.5 | 50.058 | 1600 | 11,483.4 | 8,306.0 | 55.154 |
| 820 | 5,709.4 | 4,081.0 | 50.232 | 1620 | 11,638.9 | 8,421.8 | 55.251 |
| 840 | 5,850.7 | 4,182.6 | 50.402 | 1640 | 11,794.7 | 8,537.9 | 55.347 |
| 860 | 5,992.3 | 4,284.5 | 50.569 | 1660 | 11,950.9 | 8,654.4 | 55.411 |
| 880 | 6,134.2 | 4,386.6 | 50.732 | 1680 | 12,107.5 | 8,771.2 | 55.535 |
| 900 | 6,276.4 | 4,489.1 | 50.892 | 1700 | 12,264.3 | 8,888.3 | 55.628 |
| 920 | 6,419.0 | 4,592.0 | 51.048 | 1720 | 12,421.4 | 9,005.7 | 55.720 |
| 940 | 6,561.7 | 4,695.0 | 51.202 | 1740 | 12,579.0 | 9,123.6 | 55.811 |
| 960 | 6,704.9 | 4,798.5 | 51.353 | 1760 | 12,736.7 | 9,241.6 | 55.900 |
| 980 | 6,848.4 | 4,902.3 | 51.501 | 1780 | 12,894.9 | 9,360.0 | 55.990 |
| 1000 | 6,992.2 | 5,006.3 | 51.646 | 1800 | 13,053.2 | 9,478.6 | 56.078 |
| 1020 | 7,136.4 | 5,110.8 | 51.788 | 1820 | 13,212.0 | 9,597.7 | 56.166 |
| 1040 | 7,281.0 | 5,215.7 | 51.929 | 1840 | 13,371.0 | 9,717.0 | 56.253 |
| 1060 | 7,425.9 | 5,320.9 | 52.067 | 1860 | 13,530.2 | 9,836.5 | 56.339 |

Ideal-gas properties of carbon monoxide, CO  *(Concluded)*

| $T$ R | $\bar{h}$ Btu/lbmol | $\bar{u}$ Btu/lbmol | $\bar{s}°$ Btu/lbmol·R | $T$ R | $\bar{h}$ Btu/lbmol | $\bar{u}$ Btu/lbmol | $\bar{s}°$ Btu/lbmol·R |
|---|---|---|---|---|---|---|---|
| 1900 | 13,850 | 10,077 | 56.509 | 3500 | 27,262 | 20,311 | 61.612 |
| 1940 | 14,170 | 10,318 | 56.677 | 3540 | 27,608 | 20,576 | 61.710 |
| 1980 | 14,492 | 10,560 | 56.841 | 3580 | 27,954 | 20,844 | 61.807 |
| 2020 | 14,815 | 10,803 | 57.007 | 3620 | 28,300 | 21,111 | 61.903 |
| 2060 | 15,139 | 11,048 | 57.161 | 3660 | 28,647 | 21,378 | 61.998 |
| 2100 | 15,463 | 11,293 | 57.317 | 3700 | 28,994 | 21,646 | 62.093 |
| 2140 | 15,789 | 11,539 | 57.470 | 3740 | 29,341 | 21,914 | 62.186 |
| 2180 | 16,116 | 11,787 | 57.621 | 3780 | 29,688 | 22,182 | 62.279 |
| 2220 | 16,443 | 12,035 | 57.770 | 3820 | 30,036 | 22,450 | 62.370 |
| 2260 | 16,722 | 12,284 | 57.917 | 3860 | 30,384 | 22,719 | 62.461 |
| 2300 | 17,101 | 12,534 | 58.062 | 3900 | 30,733 | 22,988 | 62.511 |
| 2340 | 17,431 | 12,784 | 58.204 | 3940 | 31,082 | 23,257 | 62.640 |
| 2380 | 17,762 | 13,035 | 58.344 | 3980 | 31,431 | 23,527 | 62.728 |
| 2420 | 18,093 | 13,287 | 58.482 | 4020 | 31,780 | 23,797 | 62.816 |
| 2460 | 18,426 | 13,541 | 58.619 | 4060 | 32,129 | 24,067 | 62.902 |
| 2500 | 18,759 | 13,794 | 58.754 | 4100 | 32,479 | 24,337 | 62.988 |
| 2540 | 19,093 | 14,048 | 58.885 | 4140 | 32,829 | 24,608 | 63.072 |
| 2580 | 19,427 | 14,303 | 59.016 | 4180 | 33,179 | 24,878 | 63.156 |
| 2620 | 19,762 | 14,559 | 59.145 | 4220 | 33,530 | 25,149 | 63.240 |
| 2660 | 20,098 | 14,815 | 59.272 | 4260 | 33,880 | 25,421 | 63.323 |
| 2700 | 20,434 | 15,072 | 59.398 | 4300 | 34,231 | 25,692 | 63.405 |
| 2740 | 20,771 | 15,330 | 59.521 | 4340 | 34,582 | 25,934 | 63.486 |
| 2780 | 21,108 | 15,588 | 59.644 | 4380 | 34,934 | 26,235 | 63.567 |
| 2820 | 21,446 | 15,846 | 59.765 | 4420 | 35,285 | 26,508 | 63.647 |
| 2860 | 21,785 | 16,105 | 59.884 | 4460 | 35,637 | 26,780 | 63.726 |
| 2900 | 22,124 | 16,365 | 60.002 | 4500 | 35,989 | 27,052 | 63.805 |
| 2940 | 22,463 | 16,225 | 60.118 | 4540 | 36,341 | 27,325 | 63.883 |
| 2980 | 22,803 | 16,885 | 60.232 | 4580 | 36,693 | 27,598 | 63.960 |
| 3020 | 23,144 | 17,146 | 60.346 | 4620 | 37,046 | 27,871 | 64.036 |
| 3060 | 23,485 | 17,408 | 60.458 | 4660 | 37,398 | 28,144 | 64.113 |
| 3100 | 23,826 | 17,670 | 60.569 | 4700 | 37,751 | 28,417 | 64.188 |
| 3140 | 24,168 | 17,932 | 60.679 | 4740 | 38,104 | 28,691 | 64.263 |
| 3180 | 24,510 | 18,195 | 60.787 | 4780 | 38,457 | 28,965 | 64.337 |
| 3220 | 24,853 | 18,458 | 60.894 | 4820 | 38,811 | 29,239 | 64.411 |
| 3260 | 25,196 | 18,722 | 61.000 | 4860 | 39,164 | 29,513 | 64.484 |
| 3300 | 25,539 | 18,986 | 61.105 | 4900 | 39,518 | 29,787 | 64.556 |
| 3340 | 25,883 | 19,250 | 61.209 | 5000 | 40,403 | 30,473 | 64.735 |
| 3380 | 26,227 | 19,515 | 61.311 | 5100 | 41,289 | 31,161 | 64.910 |
| 3420 | 26,572 | 19,780 | 61.412 | 5200 | 42,176 | 31,849 | 65.082 |
| 3460 | 26,917 | 20,045 | 61.513 | 5300 | 43,063 | 32,538 | 65.252 |

Ideal-gas properties of hydrogen, $H_2$

| T R | $\bar{h}$ Btu/lbmol | $\bar{u}$ Btu/lbmol | $\bar{s}°$ Btu/lbmol·R | T R | $\bar{h}$ Btu/lbmol | $\bar{u}$ Btu/lbmol | $\bar{s}°$ Btu/lbmol·R |
|---|---|---|---|---|---|---|---|
| 300 | 2,063.5 | 1,467.7 | 27.337 | 1400 | 9,673.8 | 6,893.6 | 37.883 |
| 320 | 2,189.4 | 1,553.9 | 27.742 | 1500 | 10,381.5 | 7,402.7 | 38.372 |
| 340 | 2,317.2 | 1,642.0 | 28.130 | 1600 | 11,092.5 | 7,915.1 | 38.830 |
| 360 | 2,446.8 | 1,731.9 | 28.501 | 1700 | 11,807.4 | 8,431.4 | 39.264 |
| 380 | 2,577.8 | 1,823.2 | 28.856 | 1800 | 12,526.8 | 8,952.2 | 39.675 |
| 400 | 2,710.2 | 1,915.8 | 29.195 | 1900 | 13,250.9 | 9,477.8 | 40.067 |
| 420 | 2,843.7 | 2,009.6 | 29.520 | 2000 | 13,980.1 | 10,008.4 | 40.441 |
| 440 | 2,978.1 | 2,104.3 | 29.833 | 2100 | 14,714.5 | 10,544.2 | 40.799 |
| 460 | 3,113.5 | 2,200.0 | 30.133 | 2200 | 15,454.4 | 11,085.5 | 41.143 |
| 480 | 3,249.4 | 2,296.2 | 20.424 | 2300 | 16,199.8 | 11,632.3 | 41.475 |
| 500 | 3,386.1 | 2,393.2 | 30.703 | 2400 | 16,950.6 | 12,184.5 | 41.794 |
| 520 | 3,523.2 | 2,490.6 | 30.972 | 2500 | 17,707.3 | 12,742.6 | 42.104 |
| 537 | 3,640.3 | 2,573.9 | 31.194 | 2600 | 18,469.7 | 13,306.4 | 42.403 |
| 540 | 3,660.9 | 2,588.5 | 31.232 | 2700 | 19,237.8 | 13,876.0 | 42.692 |
| 560 | 3,798.8 | 2,686.7 | 31.482 | 2800 | 20,011.8 | 14,451.4 | 42.973 |
| 580 | 3,937.1 | 2,785.3 | 31.724 | 2900 | 20,791.5 | 15,032.5 | 43.247 |
| 600 | 4,075.6 | 2,884.1 | 31.959 | 3000 | 21,576.9 | 15,619.3 | 43.514 |
| 620 | 4,214.3 | 2,983.1 | 32.187 | 3100 | 22,367.7 | 16,211.5 | 43.773 |
| 640 | 4,353.1 | 3,082.1 | 32.407 | 3200 | 23,164.1 | 16,809.3 | 44.026 |
| 660 | 4,492.1 | 3,181.4 | 32.621 | 3300 | 23,965.5 | 17,412.1 | 44.273 |
| 680 | 4,631.1 | 3,280.7 | 32.829 | 3400 | 24,771.9 | 18,019.9 | 44.513 |
| 700 | 4,770.2 | 3,380.1 | 33.031 | 3500 | 25,582.9 | 18,632.4 | 44.748 |
| 720 | 4,909.5 | 3,479.6 | 33.226 | 3600 | 26,398.5 | 19,249.4 | 44.978 |
| 740 | 5,048.8 | 3,579.2 | 33.417 | 3700 | 27,218.5 | 19,870.8 | 45.203 |
| 760 | 5,188.1 | 3,678.8 | 33.603 | 3800 | 28,042.8 | 20,496.5 | 45.423 |
| 780 | 5,327.6 | 3,778.6 | 33.784 | 3900 | 28,871.1 | 21,126.2 | 45.638 |
| 800 | 5,467.1 | 3,878.4 | 33.961 | 4000 | 29,703.5 | 21,760.0 | 45.849 |
| 820 | 5,606.7 | 3,978.3 | 34.134 | 4100 | 30,539.8 | 22,397.7 | 46.056 |
| 840 | 5,746.3 | 4,078.2 | 34.302 | 4200 | 31,379.8 | 23,039.2 | 46.257 |
| 860 | 5,885.9 | 4,178.0 | 34.466 | 4300 | 32,223.5 | 23,684.3 | 46.456 |
| 880 | 6,025.6 | 4,278.0 | 34.627 | 4400 | 33,070.9 | 24,333.1 | 46.651 |
| 900 | 6,165.3 | 4,378.0 | 34.784 | 4500 | 33,921.6 | 24,985.2 | 46.842 |
| 920 | 6,305.1 | 4,478.1 | 34.938 | 4600 | 34,775.7 | 25,640.7 | 47.030 |
| 940 | 6,444.9 | 4,578.1 | 35.087 | 4700 | 35,633.0 | 26,299.4 | 47.215 |
| 960 | 6,584.7 | 4,678.3 | 35.235 | 4800 | 36,493.4 | 26,961.2 | 47.396 |
| 980 | 6,724.6 | 4,778.4 | 35.379 | 4900 | 35,356.9 | 27,626.1 | 47.574 |
| 1000 | 6,864.5 | 4,878.6 | 35.520 | 5000 | 38,223.3 | 28,294.0 | 47.749 |
| 1100 | 7,564.6 | 5,380.1 | 36.188 | 5100 | 39,092.8 | 28,964.9 | 47.921 |
| 1200 | 8,265.8 | 5,882.8 | 36.798 | 5200 | 39,965.1 | 29,638.6 | 48.090 |
| 1300 | 8,968.7 | 6,387.1 | 37.360 | 5300 | 40,840.2 | 30,315.1 | 48.257 |

## TABLE A–23E

Ideal-gas properties of water vapor, $H_2O$

| $T$ R | $\bar{h}$ Btu/lbmol | $\bar{u}$ Btu/lbmol | $\bar{s}°$ Btu/lbmol·R | $T$ R | $\bar{h}$ Btu/lbmol | $\bar{u}$ Btu/lbmol | $\bar{s}°$ Btu/lbmol·R |
|---|---|---|---|---|---|---|---|
| 300 | 2,367.6 | 1,771.8 | 40.439 | 1080 | 8,768.2 | 6,623.5 | 50.854 |
| 320 | 2,526.8 | 1,891.3 | 40.952 | 1100 | 8,942.0 | 6,757.5 | 51.013 |
| 340 | 2,686.0 | 2,010.8 | 41.435 | 1120 | 9,116.4 | 6,892.2 | 51.171 |
| 360 | 2,845.1 | 2,130.2 | 41.889 | 1140 | 9,291.4 | 7,027.5 | 51.325 |
| 380 | 3,004.4 | 2,249.8 | 42.320 | 1160 | 9,467.1 | 7,163.5 | 51.478 |
| 400 | 3,163.8 | 2,369.4 | 42.728 | 1180 | 9,643.4 | 7,300.1 | 51.360 |
| 420 | 3,323.2 | 2,489.1 | 43.117 | 1200 | 9,820.4 | 7,437.4 | 51.777 |
| 440 | 3,482.7 | 2,608.9 | 43.487 | 1220 | 9,998.0 | 7,575.2 | 51.925 |
| 460 | 3,642.3 | 2,728.8 | 43.841 | 1240 | 10,176.1 | 7,713.6 | 52.070 |
| 480 | 3,802.0 | 2,848.8 | 44.182 | 1260 | 10,354.9 | 7,852.7 | 52.212 |
| 500 | 3,962.0 | 2,969.1 | 44.508 | 1280 | 10,534.4 | 7,992.5 | 52.354 |
| 520 | 4,122.0 | 3,089.4 | 44.821 | 1300 | 10,714.5 | 8,132.9 | 52.494 |
| 537 | 4,258.0 | 3,191.9 | 45.079 | 1320 | 10,895.3 | 8,274.0 | 52.631 |
| 540 | 4,282.4 | 3,210.0 | 45.124 | 1340 | 11,076.6 | 8,415.5 | 52.768 |
| 560 | 4,442.8 | 3,330.7 | 45.415 | 1360 | 11,258.7 | 8,557.9 | 52.903 |
| 580 | 4,603.7 | 3,451.9 | 45.696 | 1380 | 11,441.4 | 8,700.9 | 53.037 |
| 600 | 4,764.7 | 3,573.2 | 45.970 | 1400 | 11,624.8 | 8,844.6 | 53.168 |
| 620 | 4,926.1 | 3,694.9 | 46.235 | 1420 | 11,808.8 | 8,988.9 | 53.299 |
| 640 | 5,087.8 | 3,816.8 | 46.492 | 1440 | 11,993.4 | 9,133.8 | 53.428 |
| 660 | 5,250.0 | 3,939.3 | 46.741 | 1460 | 12,178.8 | 9,279.4 | 53.556 |
| 680 | 5,412.5 | 4,062.1 | 46.984 | 1480 | 12,364.8 | 9,425.7 | 53.682 |
| 700 | 5,575.4 | 4,185.3 | 47.219 | 1500 | 12,551.4 | 9,572.7 | 53.808 |
| 720 | 5,738.8 | 4,309.0 | 47.450 | 1520 | 12,738.8 | 9,720.3 | 53.932 |
| 740 | 5,902.6 | 4,433.1 | 47.673 | 1540 | 12,926.8 | 9,868.6 | 54.055 |
| 760 | 6,066.9 | 4,557.6 | 47.893 | 1560 | 13,115.6 | 10,017.6 | 54.117 |
| 780 | 6,231.7 | 4,682.7 | 48.106 | 1580 | 13,305.0 | 10,167.3 | 54.298 |
| 800 | 6,396.9 | 4,808.2 | 48.316 | 1600 | 13,494.4 | 10,317.6 | 54.418 |
| 820 | 6,562.6 | 4,934.2 | 48.520 | 1620 | 13,685.7 | 10,468.6 | 54.535 |
| 840 | 6,728.9 | 5,060.8 | 48.721 | 1640 | 13,877.0 | 10,620.2 | 54.653 |
| 860 | 6,895.6 | 5,187.8 | 48.916 | 1660 | 14,069.2 | 10,772.7 | 54.770 |
| 880 | 7,062.9 | 5,315.3 | 49.109 | 1680 | 14,261.9 | 10,925.6 | 54.886 |
| 900 | 7,230.9 | 5,443.6 | 49.298 | 1700 | 14,455.4 | 11,079.4 | 54.999 |
| 920 | 7,399.4 | 5,572.4 | 49.483 | 1720 | 14,649.5 | 11,233.8 | 55.113 |
| 940 | 7,568.4 | 5,701.7 | 49.665 | 1740 | 14,844.3 | 11,388.9 | 55.226 |
| 960 | 7,738.0 | 5,831.6 | 49.843 | 1760 | 15,039.8 | 11,544.7 | 55.339 |
| 980 | 7,908.2 | 5,962.0 | 50.019 | 1780 | 15,236.1 | 11,701.2 | 55.449 |
| 1000 | 8,078.2 | 6,093.0 | 50.191 | 1800 | 15,433.0 | 11,858.4 | 55.559 |
| 1020 | 8,250.4 | 6,224.8 | 50.360 | 1820 | 15,630.6 | 12,016.3 | 55.668 |
| 1040 | 8,422.4 | 6,357.1 | 50.528 | 1840 | 15,828.7 | 12,174.7 | 55.777 |
| 1060 | 8,595.0 | 6,490.0 | 50.693 | 1860 | 16,027.6 | 12,333.9 | 55.884 |

## TABLE A–23E

Ideal-gas properties of water vapor, $H_2O$ (Concluded)

| $T$ R | $\bar{h}$ Btu/lbmol | $\bar{u}$ Btu/lbmol | $\bar{s}°$ Btu/lbmol·R | $T$ R | $\bar{h}$ Btu/lbmol | $\bar{u}$ Btu/lbmol | $\bar{s}°$ Btu/lbmol·R |
|---|---|---|---|---|---|---|---|
| 1900 | 16,428 | 12,654 | 56.097 | 3500 | 34,324 | 27,373 | 62.876 |
| 1940 | 16,830 | 12,977 | 56.307 | 3540 | 34,809 | 27,779 | 63.015 |
| 1980 | 17,235 | 13,303 | 56.514 | 3580 | 35,296 | 28,187 | 63.153 |
| 2020 | 17,643 | 13,632 | 56.719 | 3620 | 35,785 | 28,596 | 63.288 |
| 2060 | 18,054 | 13,963 | 56.920 | 3660 | 36,274 | 29,006 | 63.423 |
| 2100 | 18,467 | 14,297 | 57.119 | 3700 | 36,765 | 29,418 | 63.557 |
| 2140 | 18,883 | 14,633 | 57.315 | 3740 | 37,258 | 29,831 | 63.690 |
| 2180 | 19,301 | 14,972 | 57.509 | 3780 | 37,752 | 30,245 | 63.821 |
| 2220 | 19,722 | 15,313 | 57.701 | 3820 | 38,247 | 30,661 | 63.952 |
| 2260 | 20,145 | 15,657 | 57.889 | 3860 | 38,743 | 31,077 | 64.082 |
| 2300 | 20,571 | 16,003 | 58.077 | 3900 | 39,240 | 31,495 | 64.210 |
| 2340 | 20,999 | 16,352 | 58.261 | 3940 | 39,739 | 31,915 | 64.338 |
| 2380 | 21,429 | 16,703 | 58.445 | 3980 | 40,239 | 32,335 | 64.465 |
| 2420 | 21,862 | 17,057 | 58.625 | 4020 | 40,740 | 32,757 | 64.591 |
| 2460 | 22,298 | 17,413 | 58.803 | 4060 | 41,242 | 33,179 | 64.715 |
| 2500 | 22,735 | 17,771 | 58.980 | 4100 | 41,745 | 33,603 | 64.839 |
| 2540 | 23,175 | 18,131 | 59.155 | 4140 | 42,250 | 34,028 | 64.962 |
| 2580 | 23,618 | 18,494 | 59.328 | 4180 | 42,755 | 34,454 | 65.084 |
| 2620 | 24,062 | 18,859 | 59.500 | 4220 | 43,267 | 34,881 | 65.204 |
| 2660 | 24,508 | 19,226 | 59.669 | 4260 | 43,769 | 35,310 | 65.325 |
| 2700 | 24,957 | 19,595 | 59.837 | 4300 | 44,278 | 35,739 | 65.444 |
| 2740 | 25,408 | 19,967 | 60.003 | 4340 | 44,788 | 36,169 | 65.563 |
| 2780 | 25,861 | 20,340 | 60.167 | 4380 | 45,298 | 36,600 | 65.680 |
| 2820 | 26,316 | 20,715 | 60.330 | 4420 | 45,810 | 37,032 | 65.797 |
| 2860 | 26,773 | 21,093 | 60.490 | 4460 | 46,322 | 37,465 | 65.913 |
| 2900 | 27,231 | 21,472 | 60.650 | 4500 | 46,836 | 37,900 | 66.028 |
| 2940 | 27,692 | 21,853 | 60.809 | 4540 | 47,350 | 38,334 | 66.142 |
| 2980 | 28,154 | 22,237 | 60.965 | 4580 | 47,866 | 38,770 | 66.255 |
| 3020 | 28,619 | 22,621 | 61.120 | 4620 | 48,382 | 39,207 | 66.368 |
| 3060 | 29,085 | 23,085 | 61.274 | 4660 | 48,899 | 39,645 | 66.480 |
| 3100 | 29,553 | 23,397 | 61.426 | 4700 | 49,417 | 40,083 | 66.591 |
| 3140 | 30,023 | 23,787 | 61.577 | 4740 | 49,936 | 40,523 | 66.701 |
| 3180 | 30,494 | 24,179 | 61.727 | 4780 | 50,455 | 40,963 | 66.811 |
| 3220 | 30,967 | 24,572 | 61.874 | 4820 | 50,976 | 41,404 | 66.920 |
| 3260 | 31,442 | 24,968 | 62.022 | 4860 | 51,497 | 41,856 | 67.028 |
| 3300 | 31,918 | 25,365 | 62.167 | 4900 | 52,019 | 42,288 | 67.135 |
| 3340 | 32,396 | 25,763 | 62.312 | 5000 | 53,327 | 43,398 | 67.401 |
| 3380 | 32,876 | 26,164 | 62.454 | 5100 | 54,640 | 44,512 | 67.662 |
| 3420 | 33,357 | 26,565 | 62.597 | 5200 | 55,957 | 45,631 | 67.918 |
| 3460 | 33,839 | 26,968 | 62.738 | 5300 | 57,279 | 46,754 | 68.172 |

## TABLE A–26E

Enthalpy of formation, Gibbs function of formation, and absolute entropy at 77°F, 1 atm

| Substance | Formula | $\overline{h}_f^\circ$ Btu/lbmol | $\overline{g}_f^\circ$ Btu/lbmol | $\overline{s}^\circ$ Btu/lbmol·R |
|---|---|---|---|---|
| Carbon | C(s) | 0 | 0 | 1.36 |
| Hydrogen | H$_2$(g) | 0 | 0 | 31.21 |
| Nitrogen | N$_2$(g) | 0 | 0 | 45.77 |
| Oxygen | O$_2$(g) | 0 | 0 | 49.00 |
| Carbon monoxide | CO(g) | −47,540 | −59,010 | 47.21 |
| Carbon dioxide | CO$_2$(g) | −169,300 | −169,680 | 51.07 |
| Water vapor | H$_2$O(g) | −104,040 | −98,350 | 45.11 |
| Water | H$_2$O(l) | −122,970 | −102,040 | 16.71 |
| Hydrogen peroxide | H$_2$O$_2$(g) | −58,640 | −45,430 | 55.60 |
| Ammonia | NH$_3$(g) | −19,750 | −7,140 | 45.97 |
| Methane | CH$_4$(g) | −32,210 | −21,860 | 44.49 |
| Acetylene | C$_2$H$_2$(g) | +97,540 | +87,990 | 48.00 |
| Ethylene | C$_2$H$_4$(g) | +22,490 | +29,306 | 52.54 |
| Ethane | C$_2$H$_6$(g) | −36,420 | −14,150 | 54.85 |
| Propylene | C$_3$H$_6$(g) | +8,790 | +26,980 | 63.80 |
| Propane | C$_3$H$_8$(g) | −44,680 | −10,105 | 64.51 |
| n-Butane | C$_4$H$_{10}$(g) | −54,270 | −6,760 | 74.11 |
| n-Octane | C$_8$H$_{18}$(g) | −89,680 | +7,110 | 111.55 |
| n-Octane | C$_8$H$_{18}$(l) | −107,530 | +2,840 | 86.23 |
| n-Dodecane | C$_{12}$H$_{26}$(g) | −125,190 | +21,570 | 148.86 |
| Benzene | C$_6$H$_6$(g) | +35,680 | +55,780 | 64.34 |
| Methyl alcohol | CH$_3$OH(g) | −86,540 | −69,700 | 57.29 |
| Methyl alcohol | CH$_3$OH(l) | −102,670 | −71,570 | 30.30 |
| Ethyl alcohol | C$_2$H$_5$OH(g) | −101,230 | −72,520 | 67.54 |
| Ethyl alcohol | C$_2$H$_5$OH(l) | −119,470 | −75,240 | 38.40 |
| Oxygen | O(g) | +107,210 | +99,710 | 38.47 |
| Hydrogen | H(g) | +93,780 | +87,460 | 27.39 |
| Nitrogen | N(g) | +203,340 | +195,970 | 36.61 |
| Hydroxyl | OH(g) | +16,790 | +14,750 | 43.92 |

*Source of Data:* From JANAF, *Thermochemical Tables* (Midland, MI: Dow Chemical Co., 1971), *Selected Values of Chemical Thermodynamic Properties,* NBS Technical Note 270-3, 1968; and *API Research Project 44* (Carnegie Press, 1953).

Properties of some common fuels and hydrocarbons

| Fuel (phase) | Formula | Molar mass, lbm/lbmol | Density,[1] lbm/ft$^3$ | Enthalpy of vaporization,[2] Btu/lbm | Specific heat,[1] $c_p$ Btu/lbm·°F | Higher heating value,[3] Btu/lbm | Lower heating value,[3] Btu/lbm |
|---|---|---|---|---|---|---|---|
| Carbon (s) | C | 12.011 | 125 | — | 0.169 | 14,100 | 14,100 |
| Hydrogen (g) | $H_2$ | 2.016 | — | — | 3.44 | 60,970 | 51,600 |
| Carbon monoxide (g) | CO | 28.013 | — | — | 0.251 | 4,340 | 4,340 |
| Methane (g) | $CH_4$ | 16.043 | — | 219 | 0.525 | 23,880 | 21,520 |
| Methanol (l) | $CH_4O$ | 32.042 | 49.3 | 502 | 0.604 | 9,740 | 8,570 |
| Acetylene (g) | $C_2H_2$ | 26.038 | — | — | 0.404 | 21,490 | 20,760 |
| Ethane (g) | $C_2H_6$ | 30.070 | — | 74 | 0.418 | 22,320 | 20,430 |
| Ethanol (l) | $C_2H_6O$ | 46.069 | 49.3 | 395 | 0.583 | 12,760 | 11,530 |
| Propane (l) | $C_3H_8$ | 44.097 | 31.2 | 144 | 0.662 | 21,640 | 19,930 |
| Butane (l) | $C_4H_{10}$ | 58.123 | 36.1 | 156 | 0.578 | 21,130 | 19,510 |
| 1-Pentene (l) | $C_5H_{10}$ | 70.134 | 40.0 | 156 | 0.525 | 20,540 | 19,190 |
| Isopentane (l) | $C_5H_{12}$ | 72.150 | 39.1 | — | 0.554 | 20,890 | 19,310 |
| Benzene (l) | $C_6H_6$ | 78.114 | 54.7 | 186 | 0.411 | 17,970 | 17,240 |
| Hexene (l) | $C_6H_{12}$ | 84.161 | 42.0 | 169 | 0.439 | 20,430 | 19,090 |
| Hexane (l) | $C_6H_{14}$ | 86.177 | 41.2 | 157 | 0.542 | 20,770 | 19,240 |
| Toluene (l) | $C_7H_8$ | 92.141 | 54.1 | 177 | 0.408 | 18,230 | 17,420 |
| Heptane (l) | $C_7H_{16}$ | 100.204 | 42.7 | 157 | 0.535 | 20,680 | 19,180 |
| Octane (l) | $C_8H_{18}$ | 114.231 | 43.9 | 156 | 0.533 | 20,590 | 19,100 |
| Decane (l) | $C_{10}H_{22}$ | 142.285 | 45.6 | 155 | 0.528 | 20,490 | 19,020 |
| Gasoline (l) | $C_nH_{1.87n}$ | 100–110 | 45–49 | 151 | 0.57 | 20,300 | 18,900 |
| Light diesel (l) | $C_nH_{1.8n}$ | 170 | 49–52 | 116 | 0.53 | 19,800 | 18,600 |
| Heavy diesel (l) | $C_nH_{1.7n}$ | 200 | 51–55 | 99 | 0.45 | 19,600 | 18,400 |
| Natural gas (g) | $C_nH_{3.8n}N_{0.1n}$ | 18 | — | — | 0.48 | 21,500 | 19,400 |

[1] At 1 atm and 68°F.

[2] At 77°F for liquid fuels, and 1 atm and normal boiling temperature for gaseous fuels.

[3] At 77°F. Multiply by molar mass to obtain heating values in Btu/lbmol.

Sensible heat $= \dfrac{Q_s}{Q_t}$

Enthalpy $= \dfrac{\Delta h}{\Delta \omega_s}$
Humidity ratio

Dry bulb temperature, °F

Humidity ratio, pounds moisture per pound dry air

Volume, cu. ft. per lb dry air

Enthalpy, Btu per pound of dry air

Saturation temperature, °F

10% Relative humidity

**FIGURE A–31E**

Psychrometric chart at 1 atm total pressure.

*From the American Society of Heating, Refrigerating and Air-Conditioning Engineers, Atlanta, GA.*

## Conversion Factors

| DIMENSION | METRIC | METRIC/ENGLISH |
|---|---|---|
| Acceleration | $1\ m/s^2 = 100\ cm/s^2$ | $1\ m/s^2 = 3.2808\ ft/s^2$<br>$1\ ft/s^2 = 0.3048^*\ m/s^2$ |
| Area | $1\ m^2 = 10^4\ cm^2 = 10^6\ mm^2 = 10^{-6}\ km^2$ | $1\ m^2 = 1550\ in^2 = 10.764\ ft^2$<br>$1\ ft^2 = 144\ in^2 = 0.09290304^*\ m^2$ |
| Density | $1\ g/cm^3 = 1\ kg/L = 1000\ kg/m^3$ | $1\ g/cm^3 = 62.428\ lbm/ft^3 = 0.036127\ lbm/in^3$<br>$1\ lbm/in^3 = 1728\ lbm/ft^3$<br>$1\ kg/m^3 = 0.062428\ lbm/ft^3$ |
| Energy, heat, work, internal energy, enthalpy | $1\ kJ = 1000\ J = 1000\ N{\cdot}m = 1\ kPa{\cdot}m^3$<br>$1\ kJ/kg = 1000\ m^2/s^2$<br>$1\ kWh = 3600\ kJ$<br>$1\ cal^\dagger = 4.184\ J$<br>$1\ IT\ cal^\dagger = 4.1868\ J$<br>$1\ Cal^\dagger = 4.1868\ kJ$ | $1\ kJ = 0.94782\ Btu$<br>$1\ Btu = 1.055056\ kJ$<br>$\quad = 5.40395\ psia{\cdot}ft^3 = 778.169\ lbf{\cdot}ft$<br>$1\ Btu/lbm = 25{,}037\ ft^2/s^2 = 2.326^*\ kJ/kg$<br>$1\ kJ/kg = 0.430\ Btu/lbm$<br>$1\ kWh = 3412.14\ Btu$<br>$1\ therm = 10^5\ Btu = 1.055 \times 10^5\ kJ$<br>$\quad$(natural gas) |
| Force | $1\ N = 1\ kg{\cdot}m/s^2 = 10^5\ dyne$<br>$1\ kgf = 9.80665\ N$ | $1\ N = 0.22481\ lbf$<br>$1\ lbf = 32.174\ lbm{\cdot}ft/s^2 = 4.44822\ N$ |
| Heat flux | $1\ W/cm^2 = 10^4\ W/m^2$ | $1\ W/m^2 = 0.3171\ Btu/h{\cdot}ft^2$ |
| Heat transfer coefficient | $1\ W/m^2{\cdot}°C = 1\ W/m^2{\cdot}K$ | $1\ W/m^2{\cdot}°C = 0.17612\ Btu/h{\cdot}ft^2{\cdot}°F$ |
| Length | $1\ m = 100\ cm = 1000\ mm = 10^6\ \mu m$<br>$1\ km = 1000\ m$ | $1\ m = 39.370\ in = 3.2808\ ft = 1.0926\ yd$<br>$1\ ft = 12\ in = 0.3048^*\ m$<br>$1\ mile = 5280\ ft = 1.6093\ km$<br>$1\ in = 2.54^*\ cm$ |
| Mass | $1\ kg = 1000\ g$<br>$1\ metric\ ton = 1000\ kg$ | $1\ kg = 2.2046226\ lbm$<br>$1\ lbm = 0.45359237^*\ kg$<br>$1\ ounce = 28.3495\ g$<br>$1\ slug = 32.174\ lbm = 14.5939\ kg$<br>$1\ short\ ton = 2000\ lbm = 907.1847\ kg$ |
| Power, heat transfer rate | $1\ W = 1\ J/s$<br>$1\ kW = 1000\ W = 1.341\ hp$<br>$1\ hp^\ddagger = 745.7\ W$ | $1\ kW = 3412.14\ Btu/h$<br>$\quad = 737.56\ lbf{\cdot}ft/s$<br>$1\ hp = 550\ lbf{\cdot}ft/s = 0.7068\ Btu/s$<br>$\quad = 42.41\ Btu/min = 2544.5\ Btu/h$<br>$\quad = 0.74570\ kW$<br>$1\ boiler\ hp = 33{,}475\ Btu$<br>$1\ Btu/h = 1.055056\ kJ/h$<br>$1\ ton\ of\ refrigeration = 200\ Btu/min$ |
| Pressure | $1\ Pa = 1\ N/m^2$<br>$1\ kPa = 10^3\ Pa = 10^{-3}\ MPa$<br><br>$1\ atm = 101.325\ kPa = 1.01325\ bars$<br>$\quad = 760\ mm\ Hg\ at\ 0°C$<br>$\quad = 1.03323\ kgf/cm^2$<br>$1\ mm\ Hg = 0.1333\ kPa$ | $1\ Pa = 1.4504 \times 10^{-4}\ psia$<br>$\quad = 0.020886\ lbf/ft^2$<br>$1\ psi = 144\ lbf/ft^2 = 6.894757\ kPa$<br>$1\ atm = 14.696\ psia = 29.92\ in\ Hg\ at\ 30°F$<br>$1\ in\ Hg = 3.387\ kPa$ |
| Specific heat | $1\ kJ/kg{\cdot}°C = 1\ kJ/kg{\cdot}K = 1\ J/g{\cdot}°C$ | $1\ Btu/lbm{\cdot}°F = 4.1868\ kJ/kg{\cdot}°C$<br>$1\ Btu/lbmol{\cdot}R = 4.1868\ kJ/kmol{\cdot}K$<br>$1\ kJ/kg{\cdot}°C = 0.23885\ Btu/lbm{\cdot}°F$<br>$\quad = 0.23885\ Btu/lbm{\cdot}R$ |

*Exact conversion factor between metric and English units.

†Calorie is originally defined as the amount of heat needed to raise the temperature of 1 g of water by 1°C, but it varies with temperature. The international steam table (IT) calorie (generally preferred by engineers) is exactly 4.1868 J by definition and corresponds to the specific heat of water at 15°C. The thermochemical calorie (generally preferred by physicists) is exactly 4.184 J by definition and corresponds to the specific heat of water at room temperature. The difference between the two is about 0.06 percent, which is negligible. The capitalized Calorie used by nutritionists is actually a kilocalorie (1000 IT calories).

| DIMENSION | METRIC | METRIC/ENGLISH |
|---|---|---|
| Specific volume | $1\ m^3/kg = 1000\ L/kg = 1000\ cm^3/g$ | $1\ m^3/kg = 16.02\ ft^3/lbm$<br>$1\ ft^3/lbm = 0.062428\ m^3/kg$ |
| Temperature | $T(K) = T(°C) + 273.15$<br>$\Delta T(K) = \Delta T(°C)$ | $T(R) = T(°F) = 459.67 = 1.8\ T(K)$<br>$T(°F) = 1.8T(°C) + 32$<br>$\Delta T(°F) = \Delta T(R) = 1.8\Delta T(K)$ |
| Thermal conductivity | $1\ W/m\cdot°C = 1\ W/m\cdot K$ | $1\ W/m\cdot°C = 0.57782\ Btu/h\cdot ft\cdot°F$ |
| Velocity | $1\ m/s = 3.60\ km/h$ | $1\ m/s = 3.2808\ ft/s = 2.237\ mi/h$<br>$1\ mi/h = 1.46667\ ft/s$<br>$1\ mi/h = 1.6093\ km/h$ |
| Volume | $1\ m^3 = 1000\ L = 10^6\ cm^3\ (cc)$ | $1\ m^3 = 6.1024 \times 10^4\ in^3 = 35.315\ ft^3$<br>$= 264.17\ gal\ (U.S.)$<br>$1\ U.S.\ gallon = 231\ in^3 = 3.7854\ L$<br>$1\ fl\ ounce = 29.5735\ cm^3 = 0.0295735\ L$<br>$1\ U.S.\ gallon = 128\ fl\ ounces$ |
| Volume flow rate | $1\ m^3/s = 60,000\ L/min = 10^6\ cm^3/s$ | $1\ m^3/s = 15,850\ gal/min\ (gpm) = 35.315\ ft^3/s$<br>$= 2118.9\ ft^3/min\ (cfm)$ |

[1]Mechanical horsepower. The electrical horsepower is taken to be exactly 746 W.

## Some Physical Constants

| | |
|---|---|
| Universal gas constant | $R_u = 8.31447\ kJ/kmol\cdot K$<br>$= 8.31447\ kPa\cdot m^3/kmol\cdot K$<br>$= 0.0831447\ bar\cdot m^3/kmol\cdot K$<br>$= 82.05\ L\cdot atm/kmol\cdot K$<br>$= 1.9858\ Btu/lbmol\cdot R$<br>$= 1545.37\ ft\cdot lbf/lbmol\cdot R$<br>$= 10.73\ psia\cdot ft^3/lbmol\cdot R$ |
| Standard acceleration of gravity | $g = 9.80665\ m/s^2$<br>$= 32.174\ ft/s^2$ |
| Standard atmospheric pressure | $1\ atm = 101.325\ kPa$<br>$= 1.01325\ bar$<br>$= 14.696\ psia$<br>$= 760\ mm\ Hg\ (0°C)$<br>$= 29.9213\ in\ Hg\ (32°F)$<br>$= 10.3323\ m\ H_2O\ (4°C)$ |
| Stefan–Boltzmann constant | $\sigma = 5.6704 \times 10^{-8}\ W/m^2\cdot K^4$<br>$= 0.1714 \times 10^{-8}\ Btu/h\cdot ft^2\cdot R^4$ |
| Boltzmann's constant | $k = 1.380650 \times 10^{-23}\ J/K$ |
| Speed of light in vacuum | $c_o = 2.9979 \times 10^8\ m/s$<br>$= 9.836 \times 10^8\ ft/s$ |
| Speed of sound in dry air at 0°C and 1 atm | $c = 331.36\ m/s$<br>$= 1089\ ft/s$ |
| Heat of fusion of water at 1 atm | $h_{if} = 333.7\ kJ/kg$<br>$= 143.5\ Btu/lbm$ |
| Enthalpy of vaporization of water at 1 atm | $h_{fg} = 2256.5\ kJ/kg$<br>$= 970.12\ Btu/lbm$ |